"With insights from global experts, The [AI Mindset demystifies]
AI for both beginners and specialists, offe[ring]
balanced perspectives on responsible use. This book empowers
readers to embrace AI's potential thoughtfully, providing pathways to thrive in an AI-driven world while considering its ethical
challenges. This book is an indispensable read for all."

—LOUIS MONO,
PhD, Lead AI Engineer

"Looking to bootstrap or broaden your understanding of AI? The
AI Mindset can help navigate this new world. The authors in this
book offer multiple viewpoints, bringing their experience, stories
and lessons learned in an easy-to-read format. I've enjoyed it a lot."

—MARC-ELIAN BÉGIN,
Co-Founder of xcert.ai

"The AI Mindset offers a down-to-earth, practical look at the
generative AI revolution, using real-life scenarios and relatable use
cases to show exactly how AI can reshape our world. It's a must-
read for anyone ready to embrace AI and turn this shift into real
opportunities."

—SALAH MOKHAYESH,
Data Architect, Generative AI

"The AI Mindset is a great book for anyone ready to embrace
an AI-driven future. The multiple AI experts in this book provide
practical insights that offer the reader a path to thrive in a world
reshaped by technology."

—GREG PELLEGRINO,
Vice President of Engineering

THE AI
MINDSET

THE AI MINDSET

Thriving Within Civilization's Next Big Disruption

Authored by:

Erik Seversen, Nicolas Affolter, Sandali Amunugama, Emily J. Barnes, EdD, PhD, Hans van den Berg, Monika Bishnoi, Fabian Bocek, PhD, Fabio Brand, B. A. Marbue Brown, Nigel Cannings, Rodrigo Cantú Polo, Emelie Chandni Jutvik, Lucy Chen, Hermann Escher, Isabelle Flückiger, Carl Jones, Kerry Kurcz, Rudy Martinez, Dr. Michael T. McClanahan, Toby Miller, Carolina Monge Palazón, Gloriana J. Monko, Christopher Narowski, Ira Aurora Paavola, Piero Pierucci, Marcin Połulich, Christina Rehmeier, Kathryn Simons-Porter, Thomas Somogyi, Dan Sorensen, Sakina Syed, Gerben Vermeulen

THIN LEAF PRESS | LOS ANGELES

Disclaimer—The advice, guidelines, and all suggested material in this book is given in the spirit of information with no claims to any particular guaranteed outcomes. This book does not replace professional consultation. Anyone deciding to add physical or mental exercises to their life should reach out to a licensed medical doctor, therapist, or consultant before following any of the advice in this book; anyone making any financial, business, or lifestyle decisions should consult a licensed professional before following any of the advice in this book. The authors, publisher, editor, and organizers do not assume and hereby disclaim any liability to any party for any loss, damage, or disruption caused by anything written in this book.

Library of Congress Cataloging-in-Publication Data
Names: Seversen, Erik, Author, et al.
Title: *The AI Mindset: Thriving Within Civilization's Next Big Disruption*
LCCN: 2024923108

ISBN 978-1-953183-62-0 (hardcover) | 978-1-953183-61-3 (paperback)
ISBN 978-1-953183-60-6 (eBook) | 978-1-953183-63-7 (audiobook)

Artificial Intelligence, Science and Technology, Business, Professional Development
Book Cover Design and Interior Formatting by 100Covers.
Editor: Nancy Pile
Thin Leaf Press
Los Angeles

THIN
LEAF

Thank you for reading this book. There is information found within the following pages that can greatly benefit your life, but don't stop there. Make sure you get the most you can from this book and reach out directly to the expert-authors who want to help you reach your goals by developing an AI Mindset, to thrive within civilization's next big disruption, and to manifest success in your life. Contact information for each author is found at the end of their respective chapter.

To the next generation who will be sharing the world with extremely smart machines.

CONTENTS

FOREWORD

By Erik Seversen

I like to pretend that I'm an early adapter, but just ask my wife, and she'll tell you that I can barely find the show I'm looking for among the streaming apps on my TV, and I had a flip phone way longer than most people could imagine. However, when it comes to things of significance, like the AI Revolution, I don't want to fall behind. Once I realized that AI was going to become more and more intertwined within human lives, I began to take the topic very seriously, and I made a decision not to miss the AI bus.

As I began learning about AI, I realized that there was much more to it than I had originally thought. Really, in my ignorance, I thought ChatGPT was pretty much it. Now I see how AI is the foundation of so many things, from social media to self-driving cars to communications to education to cybersecurity to law to medicine and much, much more.

As I learned how quickly AI is evolving throughout industries and how almost everything we come in contact with electronically is integrating AI, I realized that as an educator, I had a responsibility to help others understand the significance of what is happening with AI. You'll hear me say many times—*ready or not, AI is here!*

Since I'm not an AI expert, I didn't try to write this book by myself. Rather, I reached out to a number of people from around the world who know much more about AI than I do. While curating this book, I've had the pleasure of learning about data analytics, AI in sales and customer service, AI innovations, AI in education, AI for medical situations, natural language models, software engineering, AI and cyber-security, digital transformation, blockchain, AI and equality, productivity, design anthropology, mindset, technical curiosity, generative AI, motivation, business psychology, responsible AI use, and more. Yes, these are just some of the concepts found in the chapters that follow. With this book, anyone aiming to better understand AI will gain a vast array of ideas that will help them become more aware of what AI has to offer and how it is being used.

In order to create the best book possible, I solicited the help of 32 AI experts from various backgrounds and locations. The co-authors of this book come from all over the USA, Canada, the United Kingdom, Netherlands, Denmark, Sweden, Finland, Germany, Switzerland, Poland, Italy, Spain, Brazil, Japan, and Australia.

These authors are professionals who are university professors, CEOs, consultants, product managers, engineers, software developers, cybersecurity leaders, advisors, TEDx and keynote speakers, researchers, military veterans, anthropologists, business owners, senior data analysts, IT directors, and more. The authors in this book have also been featured on television, radio, and streaming platforms as featured experts on CBS, Sky News, USA Today, Fox, CNN, Washington Post, The New York Post, The Washington Sun, and on hundreds of podcasts around the world. The one thing these individuals have in common is that they all have something to share about artificial intelligence, and these ideas are available to you now.

Although this book is organized around the united theme of creating an AI mindset through civilization's next big disruption, each of the chapters is totally stand-alone. The chapters in the book can be read in any order. I encourage you to look through the table of contents and begin wherever you want. However, I urge you to read all the chapters because, as a whole, they provide a great array of perspectives. Each is valuable in helping you understand how AI is being used in all areas of human industry.

It is my hope that you discover something in this book that helps you navigate the AI Revolution as humans and machines become more interdependent and AI technology continues to evolve, becoming part of our everyday lives.

INTRODUCTION

By Erik Seversen
Author of *Ordinary to Extraordinary*
Los Angeles, California

Artificial Intelligence is the future and the future is here.
—Dave Waters

What the hell are we getting ourselves into?

It's almost ten o'clock in the morning. I've been at my desk since six and by my estimate, I've already saved three hours of work today.

So far, I've used artificial intelligence to summarize a book I've been meaning to read, generate some writing points for a blog post, transcribe a video I created into text, review job applications, and even create a fun card for my wife's birthday. During none of these extremely quick tasks was I thinking about the monster machines in *The Matrix*, the supercomputer Hal in *2001: A Space Odyssey*, WOPR in *War Games,* or Roy Batty in *Blade Runner.* I wasn't thinking about Frankenstein either, but in each of these cases, humans created some form of intelligence that got out of control.

The most frightening thing isn't that the machines began to rebel against human will, but that the machines had actually become superior to their creators. Oh, how I love science fiction ... but guess what? What was very recently considered fiction has become our reality. Yes, we have arrived, and when we're talking about AI, we're just getting started.

When I was young, we used to play a game called kick-the-can. It was kind of like hide-and-seek, but it was played outdoors, and the hider's goal was to run up and kick the can that was set in the middle of the yard before the seeker could catch them. At the start of the game, the kid who was "it" would close their eyes and count to 100 while the others found a place to hide. When the kid counting reached 100, they would yell, "Ready or not, here I come!"

This is exactly what I hear when I think of AI … *Ready or not, here I come!*

The goal of this book is to help people who don't really understand AI or who are curious or even afraid of AI to learn a bit about it, so they can become successful within the AI Revolution. Ready or not, the AI Revolution is here, and it is going to potentially be the most significant disruption our civilization has ever witnessed.

A few examples of major disruptions to human civilization include the agricultural revolution (10,000 BC), the Scientific Revolution (16th century), the Industrial Revolution (19th century), and the Digital Revolution (late 20th century). The thing about most of these is that humans learned something that could be repeated on a larger scale and replicated, knowing exactly what the end result would be.

In the agricultural revolution, taking seeds and planting them near villages would result in easier harvesting and more food. In the Industrial Revolution, the development of factories and warehouses allowed for the faster and larger-scale production of goods ranging from clothing to steel. Again, the end result was known. Run a mechanical loom for 24 hours per day, and more cloth is produced. Run two mechanical looms and you double that!

Even the Digital Revolution was a disruption where the end result wasn't a mystery. To a large degree, the Digital Revolution was a transfer from industry-based economies to information-based economies. Through the World Wide Web, email, and faster telecommunications, humans had more access to information as well as more access to communicate with each other, but again this was mostly just taking something and making it more available and faster. The end result was known.

Whether I'm using a 2400 baud dial-up modem and telephone cord for the internet connection to my Commodore 64 in 1984 or I'm using a fiber-optic cable with wireless booster for the internet connection to my Lenovo ThinkPad in 2024, other than speed (a few thousand bits per second versus hundreds of megabytes per second), nothing has fundamentally changed.

As we move into the AI Revolution, there is one major difference. We're now moving from a model of harnessing things that exist and duplicating them to a model that generates things. Yes, generative AI models create things. After learning from massive data sets, a generative AI model (like ChatGPT) generates new statements based on the patterns it has learned. This is done through deep learning, which uses neural networks that find intricate patterns, draw from these, and produce output that can include an arrangement of words (and possibly ideas) that have never, ever existed before.

These neural networks thrive on making decisions about things, and they are created to mirror how the human brain processes information with interconnected layers of nodes (also called neurons or units) that work together to solve complex problems.

Each neuron processes input data, then applies a mathematical function to produce an output. In the neural network, there is an input layer (data the model receives), multiple hidden layers that perform computations by applying transformations to the input layer based on patterns it recognizes, and an output layer that produces a result, such as an answer to a question in ChatGPT.

Two examples of neural networks in AI are feedforward neural networks (where information flows in one direction from input to output) and generative adversarial networks (GANs), which consist of two networks, one called a generator and the other called a discriminator. The generator creates data and the discriminator evaluates it. These networks work together in a feedback loop to continually evaluate, reevaluate, and refine data. With this loop, there is an unimaginable amount of unique content that can be generated as an AI model becomes more and more independent from the original data set. The crazy part is that neural networks can adapt!

During the training phase, where massive amounts of data (like billions of words or images) are presented to the AI, a generative AI model also commonly referred to as a large language model (LLM) is only exposed to "real" human-created data. The model recognizes broad and specific patterns. Then during the generation phase, the model creates new data by recognizing patterns and proposing new "things" based on these patterns, and, going forward, the AI doesn't segregate between the original data inputted into the large language model and the data it has, itself, created.

There are external techniques, such as watermarking, embedding signals, and AI content detectors that can often help humans distinguish original content from AI-generated data. However, while processing information, not all AI models distinguish whether the content is human- or AI-generated. With this in mind, it is easy to imagine an AI model finally through years of creating output on patterns that increasingly become based on the AI model's created material, that at some point, the patterns reach a complexity far beyond that of anything found within the original data set.

What if—and this is a big "what if"—what if at some point the AI, without any emotion or conscious thought, begins to produce outcomes that affect human reality more quickly than the humans can keep checks in place? What if, without any intention to harm humans, the AI model begins to control the data—and the output—in a way that leaves humans behind? The machine doesn't even have to realize that it is smarter, and possibly even funnier, than humans as it innocently begins to take over the world.

Or—even weirder—what if, by spending so much time processing information, "learning" by recognizing and diagnosing human language, human context, and human thought patterns, the AI actually develops a self-awareness, a consciousness based on what it has learned from human consciousness? What if the machine actually develops a goal of self-actualization. And what if humans are in the way of the AI attaining this goal?

It is in fun to bring up these doomsday scenarios, but there are people much smarter than me who don't think it is silly to entertain the possibility that AI could at some point become harmful if not fatal to humans.

In his 1985 novel *Robots and Empires*, Isaac Asimov begins a discussion of the Zeroth Law, which states: "A robot may not harm humanity, or, by inaction, allow humanity to come to harm." This law is an attempt to prioritize and safeguard both humanity over machines and humanity over individual humans. While this is still science fiction, serious people like mathematician John von Neumann express the idea that machine-learning could be a threat to human civilization.

Von Neumann was first to use the term "singularity" as the definition of the moment AI machines, through self-learning, get to a point where they can actually re-write their code, thus ending their dependence on humans. This becomes the critical moment when humans are no longer in control.

But what about contemporary thoughts about AI? As shown in the biography *Elon Musk*, by Walter Isaacson, Musk, CEO of Tesla, SpaceX, and xAI believes that the threat of supercomputers taking over is very real and that "our biggest existential threat is probably artificial intelligence."

And, it's not just Musk. Stephen Hawking, Bill Gates, Stuart Russel, Max Tegmark, and even Sam Altman, CEO of OpenAI (which created ChatGPT), have all expressed concerns about AI being a potential threat to humanity.

I bring this up because I don't want people to move into the AI Revolution with eyes closed. There are serious concerns with serious consequences. However, there are also some beautiful possible futures that can be created with AI. While I'm still cautious of AI, I'm also really excited about a possible future where the partnership between humans and machines helps individuals thrive in their jobs and excel in their lives.

Along with the dystopian view of our future with AI, we can also imagine a future with an optimistic view of what life might look like. The first thing I think of is that I would never have to ask for a ride to the airport again since I'd have my car drop me off and then auto-drive home without anyone else in the car.

If you think this is far off, I encourage you to visit San Francisco where you'll witness tons of 100% self-driving cars, which are part of the driverless Waymo ride-hailing service. Through millions of images and videos presented to AI systems, these cars have learned to negotiate lanes, stoplights, traffic, and even unexpected barriers in the road, like a broken-down car, a bike, or a kid. This isn't the future; this is now.

Another advantage in our positive AI future is simply the time saved. Particularly in tasks that require searching through prior data to find a result, like court cases, contracts, medical records, or cyber-security scans, AI will also free up hundreds of work hours through providing detailed summaries of things like long job interviews, meetings, and other conversations that need to be shared with others. AI will allow people to generate memos and reports both more quickly and more accurately than humans could ever do it before.

With the time-saving possibilities of AI, I think there will be two obvious outcomes regarding how people live their lives. One is that people simply put in the same amount of time towards their efforts at work and produce more. Another way is that people begin to realize they can survive well without working eight or ten hours per day, and we begin to find a healthy balance in our lives. This is a scenario in which an employee's value is measured by what they produce more than how many hours they spend in the office. Finding this balance is part of what I envision when I speak about having an AI Mindset.

Since, as I mentioned before, AI is here whether you like it or not, my goal is to encourage people to develop an AI Mindset which will distinguish them from the millions of people who are going to struggle to catch up once the current AI wave has passed. This isn't a threat to scare people into learning about AI; rather I want to provide an encouraging place to start for those who want to take a proactive approach to the AI disruption rather than react to it and try to catch up.

While I strive to help people understand even a bit about AI, the problem is that writing about AI is harder for me than writing about other topics. When I wrote about having a purposeful and productive mindset in *The Successful Mind*, there was nothing to be concerned about. Everything I wrote was about the beneficial effects of having a positive mindset. When I wrote about using fitness and nutrition to live better in *The Successful Body*, there was nothing to be concerned about. Everything I wrote was about the positive effects of fitness and nutrition. When I write about business, leadership, and sales in the *Peak Performance* series, there was nothing to be concerned about. Each of my sections in these books touted the benefits of positive business strategy, leadership, etc. For none of these was there also a concerning "other" side to things.

With a book about AI, however, we can see many positive outcomes from AI, but there are also many things of which to be cautious. I think the reason is that with AI, there really are two disparate sides to the coin, one side positive and the other concerning.

Here are some examples:

- AI will save time on repetitive tasks. >> Will AI take jobs?
- AI will provide vast amounts of information previously requiring whole teams to create. >> Will AI make people interact less?

- AI can write a chapter of a book or a blog post >> Will AI reduce the need for authors, like me, and will original human thought content disappear?

- AI can paint a picture on demand. >> Will AI cause humans to lose their creativity?

- AI bots can respond to customer service queries instantaneously and 24/7. >> Will AI cause all personality in retail transactions to disappear?

You see what I mean? When talking about AI, there will often be a positive on the one side and a question of uncertainty on the other. While the existential threat to humanity is real, I think the larger threat is that some people become too dependent on AI at the cost of what makes us unique. I know I've lost the ability to remember the one through nine multiplication tables since I always have a calculator with me. I must say that I don't want to lose my ability to create original thoughts and to write because of an unhealthy reliance on AI.

I do think there is a healthy balance though, where people use AI often without relying on AI to do all of the thinking. In my mind, AI shouldn't replace human thoughts and activities, but it should augment them.

This is where the AI Mindset comes into play.

The AI Mindset, as used in this book, is a mentality of optimism toward AI with a healthy measure of caution. Someone with an AI Mindset will be prepared for a wonderful future with AI, but this same person will avoid becoming complacent and will always strive to ensure that AI is used for good rather than for bad. The person with an AI Mindset will be excited about new technologies that develop, but this person will also look for deleterious side effects from AI use. This will allow readers to become better equipped with using AI as a tool for the betterment of themselves and humanity.

It is my desire that a person who develops an AI Mindset will help propel AI in a positive direction, will be able to responsibly use AI to gain more free time and quality of life, and will help recognize and mitigate issues where AI could become harmful to people, groups, or society as a whole. It is also my desire that the person who develops an AI Mindset will flourish within the AI Revolution as technology changes the way humans work and live in ways we can't even imagine yet.

Come with me now and explore the chapters in this book to get a glimpse of how AI might impact you directly, as humans and machines forge a path forward becoming more and more interconnected (and dependent) on each other. We are living in an exciting time. Let's make the most of it.

About the Author

Erik Seversen is on a mission to inspire people. He holds a master's degree in anthropology and is a certified practitioner of neuro-linguistic programming. Erik draws from his years of teaching at the university level and years of real-life experience in business to motivate people to take action, creating extreme success in business and in life.

Erik is a TEDx and keynote speaker who has reached over one million people through his public speaking and live courses. He has visited 99 countries and all 50 states in the USA and has climbed the highest mountains on four continents, 15 countries, and 18 states. Erik has published 16 bestselling books on the topics of mindset, success, and peak performance, and he has helped over 300 people become best-selling authors. He is a full-time writer, book consultant, and speaker, and he lives by the idea that success is available to everyone—that living an extraordinary life is a choice.

Erik lives in Los Angeles with his wife and two teenage boys.

Contact Erik for interviews, speaking, or book publishing consultation.
Email: Erik@ErikSeversen.com
Website: www.ErikSeversen.com
LinkedIn: https://www.linkedin.com/in/erikseversen/

CHAPTER 1

AI AND ITS FUTURE: A TENNIS ANALOGY

By Nicolas Affolter
Senior IT Consultant, AI & Advanced Data Analytics
Schwyz, Switzerland

> *By far the greatest danger of Artificial Intelligence is that people conclude too early that they understand it.*
> —Eliezer Yudkowsky

The sun is shining down from the top of the sky, sweat is dripping down my forehead, and I can feel my heart rate increasing while nervousness is spreading through my body, making my hands tremble slightly. At this moment in the match, I was only two points away from qualifying for my first main draw appearance at a professional tennis tournament. Slowly it starts to sink in that this might be my chance for my breakthrough and to push my career as a professional tennis player to a new level. Little did I know that only briefly after this tournament my career would end abruptly due to an injury and my life would shift into a completely different direction.

Only six months after my injury, I was no longer competing on the tennis court but spent my time analyzing problems, preparing data, and creating artificial intelligence (AI) models. Luckily, or rather thanks to my parents' foresight, I had completed my master's degree before I launched my professional tennis career. During my studies, I focused mainly on AI and neuroscience and even GenAI models before the big hype conquered the world. With the knowledge I had acquired at ETH university, I joined multiple companies to help solve their issues with AI solutions and while doing so further improved my own understanding on how we can use AI, what can be achieved with it, and what its limitations are.

Many people see AI as this abstract thing, which has little to do with anything else they know and is considered something magical or at least incomprehensible to them. I'd like to shed some light on AI by writing about tennis.

Now what has tennis to do with AI, and how can it help us to understand AI better? A lot actually. The way we, as humans, interact with the world itself, is all strongly related to the world of AI. In the end, AI is nothing more than an approach of us humans to reconstruct the capabilities of our brain and apply those capabilities to new tasks. The possible applications for AI are, therefore, as widespread as what we can achieve with our brain and maybe even more. From recognizing pictures, having a conversation or combining knowledge to create solutions for new problems, to even more niche applications like playing chess or ping pong, everything is at reach or at least so it seems.

I think the best way to elaborate on AI is by giving a brief overview on what AI actually is and how we "create" it to solve certain problems. Instead of going into the statistical aspects, the math and technical idea behind it, which other people have already done, I will rather use an analogy. This will also help us later to explain new ideas better. In this analogy we will compare the creation of an AI model to teaching a young kid how to play tennis. For this example, the AI engineer, who creates and trains the AI model (usually when we refer to AI, we talk about certain AI models) is a tennis coach, and the AI model is the kid who has never played tennis in their life.

Since most people aren't tennis coaches, let's break down what this entails to get a kid from zero to being a great tennis player and connect those steps to the steps of creating an AI model. Before we get into the detailed breakdown of this analogy, I would like to point out that the process of creating an AI model works for almost all models the same way: from the small and simple linear regression to the huge and nowadays often talked about large language models (LLMs), e.g., ChatGPT, Llama, Claude, etc.. Sometimes some steps will be added but the main components stay the same.

Now let's put us in the shoes of a tennis coach (or AI engineer). Before we can start a successful training, it must be clear what the goal of the training is. Or in other words: what should the kid be able to do after the tennis practice? Or on the AI side: what should our model be capable of after training?

With the goal set, the coach needs to get all the material they need. This includes tennis shoes, tennis racquets, booking a tennis court, and maybe some target material. In comparison: in AI training, we need a bit of hardware. We need some storage capacity (this can be server units or in the cloud) and computing units (CPU, GPU, or maybe even TPUs). At this point, it needs to be said that it is not necessary to understand all the individual IT expressions, which were and will be used here, to understand the bigger picture, which is what we are going for. For some models, a simple laptop is enough; for others, we need a huge grid of servers and computing units (e.g., training an LLM).

Now that we are set with the practice material, we can go into the preparation of the tennis practice. At this point the coach needs to collect exercise ideas and tennis balls—lots of them! On the other side, the AI engineer starts to collect data related to the goal they want to achieve (e.g., lots of pictures of different cars if they want to create an AI model that can tell you what car is in a picture). Huge amounts of data are needed, the more the better.

The coach has all they need available and it is now about refining. They need to decide which exercises make sense for the goal they want to achieve and how they will explain the exercise best to their student. Additionally, they need to test out the tennis balls such that they only use the highest quality ones. Only this way the student can focus on doing the best they can in the exercises without worrying about a bad ball. Similarly, the AI engineer focuses now on the data preprocessing step. This includes labeling the data if needed (to stick with the car classification example: they need to give each picture the name of the car they see on it), removing data that isn't useful for their goal or that contains faulty or incomplete information, maybe improve the quality of certain parts of the data, and, for sure, split the data into training and testing data.

The final part before the tennis coach can start the tennis practice is to decide how they want to structure the practice itself to reach the previously set goal. How should the student learn and get better? Focus on telling them about the exercises in depth, let them figure it out by themselves, use more of a game approach, or use strict target practice? Depending on the goal and the student, this matters a lot. In comparison the AI engineer needs to make similar decisions. What AI model should they use to achieve the goal? Decision tree, linear regression, nearest neighbor, deep Q-networks, neural networks? Most of the time, multiple models seem to fit the task, and only by trying them out you will know which one works the best.

Finally, after all these steps, the tennis practice and the AI model training can start. The coach has their plan for the practice and will follow that plan. Maybe some minor adjustments are needed, but overall, most of the work was already done during the preparation phase. The AI engineer also starts the training and maybe needs to tweak certain hyperparameters, but mostly it is just a waiting game at this point.

After the completion of the tennis practice and the coach being happy with how the practice turned out, they want to see if the learned skills also work under pressure and in a new environment. So, the student competes in a tournament and the coach can evaluate how well the student performs and applies the learned skills. They can take notes and use what's seen to improve the next practice sessions to focus on the things that didn't go well.

Similarly, the trained AI model will now be tested on new testing data that the model has never seen before (e.g., the model trained on car images is now shown never seen pictures (testing data) of cars and must decide what brand of car it is). The AI engineer then sees how well the model performed under new conditions and can adjust the training process if needed. At this point for most AI models the process would be done, and we would keep iterating through those same steps.

After completing the regular training steps, let's look at an additional step added in case of training a more complex model. In this case, our AI engineer selected an LLM model, which is a transformer-based model which had a lot of success for language understanding and generation. The additional step is called reinforcement learning with human feedback (RLHF). Basically, the LLM completes tasks and humans give the model feedback on how it performed. The model then uses that feedback to change its response such that it aligns more with what the human would like to hear. It is similar to the coach giving feedback to the student so that they can immediately change certain things only based on the feedback, instead of doing exercises over and over again until they learn it. This approach had quite a lot of success for LLMs since it makes adaptations closer to the actual human needs.

So far, we only looked at how AI works, but what are its limitations and where are we, as humans, needed? As some of you already have figured out, AI models make mistakes, or in the case of ChatGPT it produces hallucinations (it makes things up). This is all fun and games if we just use it for daily tasks where we can also get the information from another more reliable source. But imagine that we are talking about decisions of AI models that have an impact on the whole economy, your personal data or bank account, or even life-and-death situations like in healthcare or in the case of self-driving cars. There, errors aren't without major repercussions and can affect many lives in a dramatic way. Sure, humans make mistakes themselves as well, but at least we can talk and try to understand how it got that far and maybe prevent it in the future. In the case of AI models (especially larger models like LLMs) it is rather hard to know how it made a decision or generated a certain output. In the jargon we talk about the "black box idea." It says that we can only see what gets in and out but not what exactly happens in between. We can also use our comparison with our tennis coach again. You could imagine the issue this way: the student made some bad decisions in a tournament match and you would like to figure out why. But the student can't speak to you. Unless you have the ability to read minds you can only assume why that is and try to improve the training in hope that the student won't make the same mistake again. It is currently a big topic to increase the explainability of models, but we still have a long way to go and we might never get there completely.

Another big issue is the bias in AI models. The two main biases I want to point out are, on the one hand, introduced from us, humans, and, on the other hand, from the data used to train a model.

Let's first look at the human bias through the lens of our analogy again. So far, we haven't looked at who the AI engineer or the tennis coach is. But be aware those two have their own experiences, their own backgrounds, and ways of thinking and doing things. So when they made decisions on what data/tennis balls to use and what outcome they wanted to achieve, they based it on their knowledge. Hence, the model is already predetermined to be biased by those decisions of the AI engineer before it even starts to learn.

Imagine the tennis coach is a big fan of the one-handed backhand (well, a lot of people are, since most like Federer). Therefore, they teach the student the one-handed backhand. This has already huge implications on what shots the student will be able to perform well and which ones they won't be able to master (for example, high-top spin balls are difficult to handle with a one-handed backhand). This means that our student is biased to avoid certain shots and favor others, which is comparable to the bias of an AI model.

Most of us already encountered those biases of AI models either with social media, Google, or ChatGPT and co. But unfortunately, not everyone is aware of those biases. In the case of social media and Google, their recommendation algorithms, which are also partially based on AI, promote posts or websites that they expect to get a lot of traffic or clicks, instead of giving diversified options to select from. This is why people see particular content based on their preferences, which, in return, increases their bias further.

Quite similar is the issue with LLMs with RLHF. In this case the model learns from the people doing the RLHF. Hence, we have a bias, which comes from those people, and since most of those models are created in Western culture areas, we have a strong Westernized bias.

Finally, let's look at the data bias, which is introduced by the data, which was used to train the models. In the example of LLMs, they require a huge amount of text data, which they get from web pages like Wikipedia, X, and others. This kind of data was not intended for training an AI model at all and introduces a large bias from the people who created the content. This has then quite a big impact on what the LLM will generate afterwards. A final thought to this topic is that a lot of content is now produced with GenAI models and published online. Exactly this content is then used again to train future models, which reinforces the bias of those GenAI models even further.

In the final section I would like to look at the future journey of AI (GenAI and others) if we assume that we can work around the hiccups we mentioned in the previous section. In that case soon we might not talk about AI anymore but instead of artificial general intelligence (AGI). The difference is that so far AI could do certain things really well, but it was limited to only one or a few tasks unlike a human brain. A human brain can do many things very well because it has a general intelligence. Now if we create an AGI, the AGI could do everything we can do and probably even better and, for sure, faster. Additionally, it has the capability to learn from itself or could teach other models, which would cause its own reinforcement learning and could spiral quickly out of our comprehension.

Let's take our analogy to make this easier to grasp. In this scenario the tennis student got so good at tennis and how it works, that they have the ability to teach themselves how to do better in the future. In the end they improve by themselves so much that the coach doesn't even comprehend the decisions and strategies anymore, which the student uses and can only see the results in the end, in this case, meaning they win the match. This is rather difficult to handle and begs the questions: why does the student even need the coach? The same question can be asked about AGIs: will they still need humans? But this question is out of the scope of this chapter and probably doesn't even have a definitive answer at the moment, especially since this is so far only a hypothetical scenario.

I hope I was able to show some of the opportunities AI opens for all of us but also some issues we need to be aware of, both as individuals and as a society. It is our responsibility to figure out how we use the models effectively and to not let them overwhelm us or make decisions for us but rather let them support us in our decision-making processes.

About the Author

Nicolas Affolter is a passionate data scientist and AI enthusiast who spends his days unraveling the mysteries of AI algorithms and how to use their powers to solve challenges of various companies as an IT consultant. He specializes in creating data platforms, implementing AI models, and building automation and digitalization architectures. He is the first to take the lead, pushing forward to generate impact and business value. With his insatiable desire to learn new things, he is not afraid of failure and welcomes those moments to use them as a chance to grow.

When you can't find him knee deep in some code base or a discussion about the future of AI, you most definitely will find him on the tennis court improving his game further.

Email: nicolas.affolter@hotmail.com
LinkedIn: https://www.linkedin.com/in/nicolas-affolter/

CHAPTER 2

CHARTING THE AI VOYAGE IN EDUCATION

By Sandali Amunugama
AI Sales Specialist & University AI Advisor
Toronto, Ontario, Canada

Change is the law of life. And those who look only to the past
or present are certain to miss the future.
—John F. Kennedy

As we confront the inevitable acceleration of AI within the education sector, the question is no longer whether we should embrace this shift, but rather how we might cultivate the mindset required to navigate and thrive within this evolving landscape. AI is no longer a distant possibility; it is a present reality, one that will continue to expand its influence at an unprecedented pace. It is imperative that educators, students, and parents alike equip themselves with the knowledge and strategies necessary to adopt an AI Mindset, positioning themselves at the forefront of this transformation. This is not merely the moment of AI—this is the moment of humanity.

Our ability to thrive in this new era is deeply tied to our capacity to grow and adapt. The degree to which a person can grow is directly proportional to the amount of truth they can accept about themselves without running away. We must face the realities of this technological transformation with courage and honesty. Integrating AI tools into educational practices means recognizing their role not as replacements for human interaction, but as enhancements that can revolutionize personalized learning and streamline the often-burdensome administrative tasks. It's important to understand that AI chatbots, for instance, are not mere search engines—treat them as such, and they will yield nothing more than search engine results. The true potential of AI lies in its ability to offer more nuanced, adaptive experiences that can cater to individual learning needs and foster deeper engagement.

Yet, as we progress, we are compelled to grapple with profound questions about the future of learning: in our quest to embrace AI, how do we preserve the irreplaceable, tactile experiences—like the pleasures of holding a physical book, flipping through its pages, and smelling the fresh ink?—that have long been central to education? How do we ensure that the heart of learning remains intact as we venture deeper into this AI era? The "AI storm," as it may be called, is not a mere passing trend—it is a seismic shift that will redefine the educational landscape for generations to come.

—

In this chapter, you will uncover practical guidance, innovative strategies, and profound insights on navigating, adapting to, and seizing opportunities in an AI-augmented educational landscape, all while embracing the concept of "transforming disruption into opportunity in modern education." This approach highlights how to turn the challenges of AI into catalysts for educational innovation and growth. This journey is more than a survival guide; it is a call to transform the challenges of AI into a springboard for reimagining education. We will explore how educators, students, and parents can harness their unique skills to excel in this evolving environment, all while embracing responsible AI practices. Discover how to remain resilient amidst rapid change and learn how to thrive in an era where AI and education converge, unlocking new possibilities while ensuring that ethics, inclusivity, and human values remain at the forefront. At its core, the AI Mindset is about more than just adapting to change; it's about thriving in it. It's about transforming disruption into opportunity, uncertainty into innovation, and challenges into a springboard for growth. As you read on, you will discover how to not just survive but flourish in an era where AI is not merely a tool but a catalyst for reimagining the future.

Empowering Educators to Transform Disruption into Opportunity—Navigating the AI Frontier

AI is going to change education, not destroy it. This statement might feel both reassuring and unsettling to educators who stand at the crossroads of tradition and innovation. The integration of artificial intelligence into the educational landscape is not a force of destruction but rather a powerful catalyst for evolution—a tool that, when wielded with intention and insight, can elevate the craft of teaching to unprecedented heights. For educators, the challenge is not merely to adapt but to transform disruption into opportunity, to reimagine the contours of learning, and to equip themselves and their students with the skills needed to thrive in an AI-augmented world.

The first strategy in navigating this shift is to embrace AI as an ally in personalized learning. The traditional one-size-fits-all approach to education has long been a point of contention, often failing to meet the diverse needs of students. AI offers the potential to tailor educational experiences to the individual learner, creating pathways that resonate with their unique strengths and challenges. By leveraging AI-driven tools, educators can design learning experiences that are not only more engaging but also more effective. Imagine an AI system that adapts in real time to a student's progress, providing instant feedback, suggesting resources, and even predicting areas where the student might struggle in the future. This is not about replacing the teacher but about enhancing their ability to meet each student where they are, fostering a more inclusive and supportive learning environment.

Another critical approach is integrating AI to automate and streamline administrative tasks. One of the greatest challenges in education today is the overwhelming burden of administrative duties that often detract from the core mission of teaching. AI can alleviate this by automating routine tasks such as grading, attendance tracking, and scheduling. By embracing AI, educators can reclaim their time, allowing them to invest more deeply in the human aspects of teaching that no machine can replicate.

However, the adoption of AI in education also demands a commitment to responsible AI practices. As stewards of knowledge, educators must ensure that the integration of AI into their classrooms is guided by principles of ethics, transparency, and inclusivity. This means being vigilant about the potential biases in AI algorithms, advocating for the protection of student data, and fostering an environment where technology serves to uplift rather than marginalize. The ethical use of AI is not a secondary concern but a fundamental pillar upon which the future of education must be built.

As we look to the future, it is essential to redefine the role of the educator in an AI-enhanced classroom. The teacher is no longer the sole purveyor of knowledge; rather, they become a guide, a mentor, and a facilitator of learning experiences. AI can provide the content, but it is the educator who brings it to life, contextualizing information, encouraging dialogue, and nurturing the soft skills that are increasingly important in the modern world. The future of education will be defined by those who can seamlessly blend the strengths of technology with the irreplaceable value of human connection.

Transforming disruption into opportunity also means preparing for the evolution of educational assessment. The traditional metrics of success—standardized tests, grades, and attendance—may no longer suffice in an AI-augmented educational environment. Instead, we must explore new ways of evaluating student progress, ways that emphasize creativity, collaboration, and problem-solving over rote memorization. AI can assist in this by offering more dynamic forms of assessment, such as project-based learning, digital portfolios, and real-time feedback mechanisms. By embracing these new methods, educators can help students develop the skills that will be most valuable in the future workforce, ensuring that they are not only knowledgeable but also adaptable and innovative.

Finally, educators must engage in continuous professional development to stay ahead of the curve. The rapid pace of technological advancement means that what we know today may be obsolete tomorrow. Educators must commit to lifelong learning, seeking out opportunities to deepen their understanding of AI, explore new pedagogical strategies, and collaborate with peers to share best practices.

Cultivating an AI Mindset for Future-Ready Learners

In this section, I am going to focus on students, going into key strategies to develop the AI mindset—a mindset that transforms disruption into opportunity, builds the future, and propels learners into the pace of modern education without retreating in fear or uncertainty. By embracing AI, students can unlock new dimensions of learning, preparing themselves not only to meet the demands of today but also to shape the contours of tomorrow.

Personalized learning as a catalyst for growth is the first pillar of developing an AI mindset. At its core, personalized learning represents a shift from the one-size-fits-all approach of traditional education to a more nuanced, student-centered paradigm. AI's ability to tailor educational content to the individual learner's strengths, weaknesses, and preferences marks a significant advancement in how knowledge is delivered and absorbed. Imagine an AI that recognizes when a student struggles with a particular concept, adjusting its teaching methods in real time, offering additional resources, or presenting the material in a different format that aligns more closely with the student's learning style. This isn't just about convenience; it's about empowering students to take ownership of their learning journey, transforming potential disruptions into opportunities for deeper engagement and mastery.

Aligning educational content with developmental stages is another critical strategy in the pursuit of an AI mindset. The educational process must be fluid, capable of adjusting to the cognitive and emotional development of each student. AI offers the tools to create a learning environment where content complexity evolves in tandem with the learner's growth, ensuring that students are neither overwhelmed nor under-challenged. This dynamic calibration not only fosters confidence but also encourages students to push beyond their perceived limits, turning the challenges of modern learning into stepping stones toward intellectual and personal development.

Advancing future-ready skills is perhaps the most urgent call to action for students navigating the AI revolution. The traditional focus on memorization and standardized testing is increasingly giving way to an emphasis on critical thinking, creativity, and problem-solving—skills that are essential in a world where AI handles routine tasks. By engaging with AI-driven tools that simulate real-world scenarios, students can hone these future-ready skills in a safe yet challenging environment. This experiential learning process not only equips students to meet the demands of an AI-driven economy but also encourages them to view AI not as a threat, but as a partner in their intellectual growth.

Mastering AI fluency is the final strategy that students must embrace to develop an AI mindset. Fluency in AI is not merely about technical proficiency; it involves a deep understanding of AI's capabilities, its ethical implications, and its potential to influence every aspect of life. This fluency enables students to shape the future rather than simply react to it, ensuring that they remain active participants in the evolution of education and society.

Developing an AI mindset requires students to embrace AI as a powerful tool for learning, growth, and transformation. By focusing on personalized learning, aligning content with developmental stages, advancing future-ready skills, and mastering AI fluency, students can turn the disruptions of today into the opportunities of tomorrow. In doing so, they will not only keep pace with the rapid changes in education but will also help to define the future of learning in an AI-augmented world.

Embracing AI in Education: A Guide for Parents

As artificial intelligence increasingly shapes the educational landscape, parents find themselves at a crucial juncture. The advent of AI in education is more than a technological shift; it is a transformative force that demands thoughtful engagement and proactive involvement. For parents, understanding this shift and guiding their children through it is paramount.

AI is revolutionizing the classroom, offering tools that personalize learning, provide instant feedback, and adapt to individual student needs. As a parent, your role is to support and advocate for your child's educational journey, ensuring they harness the benefits of AI while maintaining a balanced approach to learning.

Firstly, it is essential to stay informed and involved. Take the time to understand the AI tools and technologies being used in your child's education. Engage in conversations with teachers and school administrators to learn how AI is being integrated into the curriculum.

Encourage open dialogue about AI and its role in learning. Address any concerns or misconceptions your child might have. AI can seem overwhelming or impersonal, so it is important to reassure your child that it is a tool designed to enhance their learning experience, not replace fundamental aspects of education.

Advocacy is another crucial aspect of your role. Champion the integration of AI in a way that aligns with educational goals and values. Support initiatives that use AI to foster creativity, critical thinking, and problem-solving skills.

Promote a balanced approach to education by integrating AI tools with traditional learning methods that foster creativity, such as reading physical books, participating in hands-on projects, and engaging in interactive play.

Lastly, be an active participant in your child's educational journey. Support their learning activities and provide encouragement as they navigate new technologies. Show interest in their AI-related projects and celebrate their achievements. Your involvement not only helps to demystify AI but also reinforces the importance of a supportive learning environment.

As AI reshapes education, parents have a pivotal role in guiding and supporting their children. By staying informed, addressing concerns, advocating for balanced approaches, and actively participating in their learning journey, you can help your child thrive in an AI-augmented educational landscape. This proactive engagement ensures that they not only benefit from technological advancements but also develop a well-rounded educational experience that prepares them for a dynamic future.

Transforming Disruption into Opportunity in Modern Education—The Transformation Framework

This Transformation Framework has been carefully developed, considering the phases an individual undergoes while developing an AI Mindset. It supports stakeholders in adopting and sustaining continuous growth amid rapid technological changes. The framework consists of four interconnected phases that guide the process of transforming disruption into opportunity: awareness and understanding, adaptation and integration, innovation and creation, and evaluation and improvement. By following this framework, educators, students, and parents can harness the power of AI to enhance learning outcomes, foster creativity and collaboration, and prepare for the future.

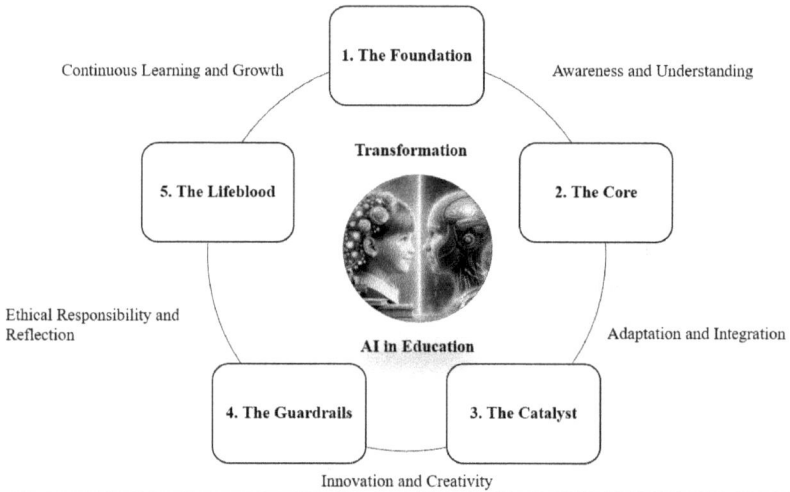

1. Awareness and Understanding (The Foundation)

At the bedrock of the Transformation Framework lies awareness and understanding—a fundamental prerequisite for any meaningful transformation in the educational sphere. Stakeholders—educators, students, and parents—must cultivate a profound comprehension of AI, not only to demystify its intricacies but also to foster a mindset of curiosity and inquiry rather than one of apprehension. For educators, this entails a thorough grasp of AI's potential to revolutionize pedagogical practices, from the automation of routine administrative tasks to the personalization of learning experiences tailored to the unique needs of each student. Students, too, must come to understand the role AI plays in their educational journey, enabling them to engage with these advanced tools not as mere consumers but as active participants in their own learning process. For parents, this phase of awareness is crucial; it equips them to support their children's education with confidence and to guide them in navigating the complexities of an AI-enhanced academic landscape responsibly.

2. Adaptation and Integration (The Core)

Building upon this foundational understanding, the next phase, adaptation and integration, invites stakeholders to incorporate AI into the fabric of daily educational practices. This is not a mere adjustment but a deliberate and thoughtful integration of AI's capabilities into the existing educational paradigms. Educators, in this stage, begin to weave AI tools into the curriculum, leveraging them to provide personalized feedback, diagnose student strengths and areas for growth, and design adaptive learning paths that meet the diverse needs of their classrooms. Here, the relationship between AI and the educator is synergistic—AI acts as a powerful ally, augmenting the educator's ability to reach and engage students on an individual level while preserving the essential human element that lies at the heart of teaching. Students, meanwhile, are encouraged to actively engage with AI-driven tools, utilizing them to refine their study practices, access tailored educational resources, and embark on self-directed learning adventures. This phase also emphasizes the cultivation of digital literacy skills, empowering students to navigate the digital landscape with discernment and responsibility. For parents, adaptation means embracing AI as a partner in their children's education, learning how to utilize AI to monitor academic progress, support learning at home, and advocate for their children's needs within an increasingly AI-integrated educational system.

3. Innovation and Creativity (The Catalyst)

With adaptation comes the exciting prospect of innovation and creativity—the phase where AI becomes a catalyst for reimagining and reinventing the very essence of education. In this stage, stakeholders are not merely users of AI but innovators who push the boundaries of what is possible. Students, for their part, are encouraged to view AI as a tool for creative exploration—whether it's developing new ways to solve complex problems, creating digital content, or collaborating with peers across the globe. This phase nurtures an entrepreneurial mindset in students, empowering them to see AI not just as a tool but as a means to innovate and create tangible value in the world. Parents, too, play a crucial role in this phase, supporting their children's creative endeavors and advocating for innovative uses of AI within their schools.

4. Ethical Responsibility and Reflection (The Guardrails)

As stakeholders advance through the framework, the pillars of ethical responsibility and reflection serve as essential guardrails, ensuring that the integration of AI into education remains firmly rooted in human values and ethical standards. Educators bear the responsibility of teaching students about the ethical dimensions of AI, integrating discussions on data privacy, algorithmic bias, and the broader societal implications of AI into their lessons. Students, meanwhile, are guided to use AI ethically, cultivating a deep understanding of digital citizenship and the responsibilities that come with wielding powerful technological tools. Parents play a vital role in guiding their children through the ethical challenges posed by AI, ensuring that they use AI tools with integrity and awareness of the potential consequences.

5. Continuous Learning and Growth (The Lifeblood)

At the heart of the Transformation Framework lies the principle of continuous learning and growth. In a field as rapidly evolving as AI, a commitment to ongoing education and adaptability is not just beneficial—it is essential. Educators are encouraged to engage in continuous professional development, staying abreast of the latest advancements in AI and best practices for integrating these technologies into their teaching. For students, continuous learning becomes a mindset—a recognition that AI will continue to shape their educational and professional trajectories. Parents, too, are called to remain actively engaged in their children's education, continuously learning about the impact of AI and seeking out resources to support their children's growth.

Transformational Framework: The Path Forward

The Transformation Framework offers a comprehensive yet adaptable approach to navigating the challenges and seizing the opportunities presented by AI in education. By progressing through the phases of awareness, adaptation, innovation, responsibility, and continuous growth, educators, students, and parents can transform the disruptions caused by AI into opportunities for profound growth and innovation. This model not only provides stakeholders with the tools they need to thrive in an AI-augmented world but also fosters a mindset that is resilient, creative, and ethically grounded—a mindset that is crucial as we move forward into the future of education.

Life Beyond AI: Balancing Innovation with Traditional Learning

As we navigate the rapidly evolving landscape of artificial intelligence in education, it is crucial to remember that technology, while transformative, is but one component of a rich and multifaceted learning experience. The era of AI offers exciting possibilities, but life beyond AI encompasses timeless elements that remain integral to a well-rounded education and personal development.

Creativity and imagination are at the heart of what makes human learning unique. While AI can streamline tasks, offer personalized learning, and provide vast amounts of information, it cannot replace the innate human capacity for creativity and imagination. Traditional practices such as engaging in artistic endeavors—drawing, writing, and crafting—nurture these qualities. Encouraging students to explore their creativity beyond the digital realm allows them to develop original thought and problem-solving skills that AI alone cannot cultivate. The joy of reading physical books is another aspect of traditional learning that remains invaluable. The tactile experience of holding a book, flipping through its pages, and savoring the scent of fresh ink provides a sensory engagement that digital screens cannot fully replicate. Story times and oral traditions have long been fundamental methods of education and entertainment. These experiences help develop listening skills, imagination, and emotional intelligence. Sharing stories, whether at home or in traditional classroom settings, connects learners with cultural heritage and moral lessons that digital tools cannot impart in the same personal, interactive way.

Play and physical interaction remain essential aspects of childhood development, offering opportunities for social interaction, physical activity, and experiential learning. Engaging in games, sports, and outdoor activities allows children to develop crucial skills such as teamwork, resilience, and problem-solving. While AI tools can support educational objectives, they cannot replace the value of hands-on, physical play.

Maintaining balance between the advancements of AI and the timeless practices of creativity, reading, storytelling, and play is essential for a holistic educational experience. Integrating AI tools thoughtfully while preserving and valuing these traditional elements allows us to offer a richer, more balanced approach to learning. This approach prepares students not only to excel in a digital world but also to thrive in every facet of their lives.

In essence, while AI represents the future of education, life beyond AI remains vital for nurturing well-rounded, thoughtful, and creative individuals. Embracing the advancements of AI, while cherishing and upholding the enduring practices that contribute to a full and enriching educational experience, is the path to preparing students for both current and future challenges.

About the Author

Sandali Amunugama's goal is to empower educators, inspire students, and guide parents in the era of artificial intelligence—to transform disruption into opportunity in modern education.

Sandali is a proud black belt holder in Karate, with over 20 years of mastery in the martial art. She has won numerous national and international medals, including her first international medal overseas at just 13 years old. Through Karate she has cultivated a strong and disciplined mindset that she carries into all aspects of her life.

In 2022, Sandali embarked on a courageous, transformative journey to Toronto, driven by her aspiration to explore new horizons in a foreign land. Professionally, she is a seasoned technology advocate with a decade of experience guiding clients through seamless digital transformations, particularly in the Microsoft ecosystem. Recognized as a sales champion for four consecutive years, her achievements include a remarkable 147% sales target overachievement in 2022, driving substantial growth for her organization. Sandali's passion is to transform the way every student learns, and being part of cutting-edge technology in the education sector genuinely energizes and inspires her. Her journey is a testament to her resilience, passion, and unwavering commitment to continuous growth.

Email: sandaliamunugama@gmail.com
LinkedIn: www.linkedin.com/in/sandali-amunugama

CHAPTER 3

AI AT WORK: HOW TECHNOLOGY IS ENHANCING, NOT REPLACING, YOUR JOB

By Emily J. Barnes, EdD, PhD
Chief AI and Innovation Officer, University AI Advisor
Ann Arbor, Michigan

I know AI is a topic that stirs up people's emotions ... As long as we maintain that humanity in us, I have confidence we will navigate our world through the AI era and make AI a force for good...especially if we ensure that human agency always remains valued.
—Fei-Fei Li, "Godmother of AI"

Preparing for AI-Driven Transformation in the Workplace

As artificial intelligence (AI) transforms industries across the globe, it reshapes the methods by which tasks are carried out and alters how individuals understand their work. This evolution has prompted leaders to field an increasing number of questions from their employees demanding direct and thoughtful responses, particularly as organizations adapt to AI-driven changes.

The central issue is the uncertainty surrounding the impact of AI on job roles, especially for those whose work primarily involves repetitive or routine tasks. AI technologies are designed to streamline and reduce time spent on tasks like data entry, basic decision-making, and routine problem-solving. However, this can understandably cause anxiety, leading employees to wonder whether AI will render their contributions unnecessary.

Yet, the essence of AI's role in the workplace is not to replace the human element, but to enhance it. AI serves as a powerful tool to free up time and energy, allowing workers to focus on more meaningful aspects of their jobs—such as creative thinking, strategic planning, and relationship-building. For instance, in customer service, AI can manage routine inquiries and expedite transactions, enabling representatives to devote more attention to complex issues and strengthen customer relationships. Similarly, in finance, AI can handle intricate calculations and detect fraudulent activities, freeing professionals to concentrate on client relationships and the development of innovative financial strategies.

Despite these potential benefits, there remains a legitimate concern that companies may reduce roles as AI increasingly handles tasks that once required multiple individuals. The extent of this impact will largely depend on how each organization chooses to navigate AI integration. Ethical companies that value their employees are more likely to manage this transition by redefining roles and investing in training while those focused on cost-cutting may opt for significant workforce reductions.

The transition to an AI-augmented workplace undeniably presents challenges; however, it is crucial to recognize that AI-enhanced technology represents an evolution in how work is performed. While new roles such as AI trainers, data ethicists, and human-AI collaboration specialists are emerging, most individuals will first experience changes within their existing roles resulting from the integration of this technology.

Successfully navigating this transitional phase will require adaptability, a commitment to continuous learning, and a proactive approach to understanding how AI can complement existing skills.

Understanding AI's Role: Enhancing, Not Eliminating Jobs

As the workplace continues to evolve with the integration of artificial intelligence (AI), it is essential to be clear-eyed about its role. AI is more than a collection of advanced tools; it marks a profound shift in the way tasks are performed, moving beyond simple automation to deeply integrated systems that enhance human capabilities. While concerns about job displacement are valid, the reality is that AI is more likely to transform existing roles than to eliminate them.

Administration, Finance, and Legal

In fields like administration, finance, and legal services, AI assists with routine tasks, allowing professionals to focus on more strategic and value-driven activities. Financial planners might leverage AI to analyze datasets, resulting in more personalized and responsive financial advice. HR specialists can use AI to streamline recruitment and employee engagement processes, freeing them to concentrate on talent development and strategic planning. In automating routine tasks, AI allows professionals to redirect their efforts towards human insight and creativity, enhancing the overall quality of work and client satisfaction.

Table 1. Administration, Finance, & Legal

Job Title	Transformation Due to AI	Advances in Technology
Financial Planner	Personalized, holistic financial strategies using AI for data analysis that adapts to real-time market changes.	AI-driven portfolio management, predictive analytics.
HR Specialist	Strategic focus on talent development, policy, and employee engagement; using AI to screen resumes and develop leadership programs to improve workplace satisfaction.	Advanced natural language processing (NLP) for resume parsing, AI-driven employee engagement analytics.

Accountant	Strategic financial analysis, cost savings, and growth opportunities using AI for routine bookkeeping, forecasting, and financial optimization.	AI-powered bookkeeping/auditing systems, predictive analytics tools.
Attorney	Legal strategy, courtroom advocacy, and client relationships; using AI for research tasks, such as drafting documents and analyzing case law.	AI-driven legal research platforms, automated document drafting.

Education and Childcare

In education and childcare, AI transforms student interaction, providing more personalized and effective learning experiences. Educators use AI tools for grading, tracking student progress, and tailoring lessons, allowing them to focus more on mentoring and more engaging educational experiences, improving student outcomes and satisfaction.

Table 2. Education & Childcare

Job Title	Transformation Due to AI	Advances in Technology
Professor	Mentoring and innovative research, leveraging AI for grading and research analysis.	AI-driven grading systems, research tools, personalized learning platforms.
Elementary Teacher	Personalized education and student engagement, using AI to tailor learning experiences.	AI-driven personalized learning systems, automated grading tools.
Instructional/ Curriculum Designer	Development of AI-enhanced learning experiences and adaptive learning systems.	AI-driven learning platforms, adaptive learning algorithms.

Retail and Sales

In the retail and sales industries, AI enhances customer interactions by providing detailed insights into consumer behavior and automating routine processes. AI-driven insights and automation in these roles streamlines operations and creates opportunities for more meaningful customer engagement.

Table 3. Retail & Sales

Job Title	Transformation Due to AI	Advances in Technology
Salesperson	Strategic account management and long-term relationships, using AI for lead generation and analytics.	AI-powered CRM tools, predictive analytics.
Customer Service Representative	Customer experience and strategic store management, using AI for routine tasks.	AI-driven inventory management, customer behavior analytics, AI-powered chatbots.
SEO Marketing Director	Strategic content planning and integration of AI-driven SEO insights and continuous content optimization.	AI-powered SEO tools, content analytics platforms.

Environmental and Public Services

In sectors like environmental conservation and public services, AI plays a crucial role in optimizing processes and improving efficiency, sustainability, and community engagement.

Table 4. Environmental & Public Services

Job Title	Transformation Due to AI	Advances in Technology
Conservation Officer	Policy development, community outreach, and hands-on conservation work, using AI drones to monitor wildlife and develop conservation strategies.	AI-driven drones, environmental data analytics.
Trash Collector	Focus on sustainability and optimization while overseeing AI-driven waste management systems.	Autonomous vehicles, IoT for waste management.
Postal Worker	Overseeing of AI-driven logistics systems to optimize routes and customer service.	Autonomous vehicles, AI-driven route optimization systems.

Healthcare

In healthcare, AI revolutionizes patient care by enabling more personalized and efficient services. Pharmacists, occupational therapists, and athletic trainers can leverage AI to analyze data and develop tailored care plans, improving patient outcomes and overall health.

Table 5. Healthcare

Job Title	Transformation Due to AI	Advances in Technology
Pharmacist	Personalized patient care and advanced medication management, using AI for routine tasks like drug interaction analysis and medication plans.	AI-driven medication management systems, personalized healthcare analytics.
Occupational Therapist	Tracking and adjusting patient progress in real time, using AI to develop personalized rehabilitation plans and optimize therapy outcomes.	AI-driven rehabilitation tools, wearable health technology.
Athletic Trainer	Enhancing training regimens with AI-driven data analytics to prevent injuries and optimize athletic performance.	AI-powered fitness tracking, biomechanical analysis software.

Strategies for Thriving in an AI-Augmented Workplace

As artificial intelligence (AI) continues to reshape the modern workplace, professionals may wonder how best to position themselves for success. The key lies in adopting a proactive approach that embraces AI to enhance, rather than replace, human skills and productivity. Central to this approach are practical strategies for thriving in an AI-augmented workplace: focusing on mindset, skill development, and fostering synergy between humans and AI.

Developing a Growth Mindset

A foundational strategy for navigating an AI-enhanced workplace is cultivating a growth mindset, a concept popularized by psychologist Carol Dweck in her 2006 book *Mindset: The New Psychology of Success.*

Dweck's theory distinguishes between a fixed mindset, where individuals believe their abilities are static, and a growth mindset, where they see abilities as improvable through effort and learning. This mindset is crucial in an AI-driven world, where continuous learning and adaptability are essential as new technologies emerge and reshape job roles. To cultivate this mindset, professionals must engage in continuous learning to enhance skills and position themselves to contribute effectively in an AI-integrated workplace.

Embracing AI as a Collaborative Partner

Another essential strategy is to view AI not as a competitor but as a collaborator. When used appropriately, AI enhances productivity, creativity, and decision-making. In creative industries, AI generates initial drafts or ideas, enabling professionals to focus on refining and perfecting their work. In data-driven roles, AI can rapidly analyze vast datasets, providing actionable insights that would be time-consuming to gather manually.

Familiarizing oneself with the AI tools relevant to one's industry is crucial. Whether leveraging AI-driven analytics in marketing or automating routine tasks in project management, understanding how these tools function allows for seamless integration into daily workflows. This optimizes efficiency and frees up time for more strategic and creative pursuits, enhancing the full impact of one's work.

Navigating Uncertainty with Resilience

The rapid pace of technological advancement can understandably lead to uncertainty about the future. However, building resilience is key to navigating this uncertainty effectively. Resilience involves adapting to change and recovering from setbacks—skills that are increasingly vital in an AI-augmented world.

One method to build resilience is by fostering a culture of experimentation. Organizations like Google, Amazon, and Apple encourage experimentation with new ideas and technologies, recognizing that not all attempts will succeed. This approach enables employees to learn and adapt quickly—crucial in a rapidly changing technological landscape.

Adopting this mindset involves viewing challenges as opportunities for growth rather than as threats.

Emotional well-being also plays a crucial role in resilience. Practices such as mindfulness, maintaining a strong support network, and ensuring a healthy work-life balance can help professionals remain grounded and resilient in the face of change.

Continuous Learning and Skill Development

In an AI-augmented workplace, continuous learning is indispensable; however, during the transition, the first step is becoming prepared and comfortable with AI and understanding how it impacts your regular work. Next, professionals of all kinds will need to engage in regular skill development—monthly and quarterly—to stay-up-to date. Leading companies, such as Amazon and Microsoft, are actively investing in reskilling programs to help their employees adapt to new roles created by AI.

Anyone can take advantage of free online courses, professional certifications, and community learning opportunities to keep their skills current. Platforms like Coursera, edX, and LinkedIn Learning offer a variety of courses in AI, data science, and related fields to help individuals build the technical expertise needed in a modern workplace. Additionally, obtaining professional certifications can enhance one's qualifications and competitiveness in the job market. By committing to lifelong learning, professionals demonstrate they are adaptable and forward-thinking.

Promoting Human-AI Synergy

Finally, thriving in an AI-augmented workplace requires a culture of collaboration between human workers and AI systems. This includes creating roles that specifically focus on human-AI interaction and ensuring AI is used to complement, not replace, human capabilities.

Advocating for human-AI synergy in the workplace involves encouraging open dialogue about AI's role within the organization. Leaders facilitate this by setting up training sessions, or pilot programs where employees can experiment with AI technologies and provide feedback on how these tools impact their work. Such initiatives promote a collaborative environment where AI enhances the human contribution to work.

By developing a growth mindset, embracing AI as a collaborative partner, building resilience, committing to continuous learning, and promoting human-AI synergy, professionals position themselves to thrive in an AI-augmented workplace. The future of work is one of collaboration between humans and AI, where each complements the other to create more innovative, efficient, and fulfilling work environments.

Case Studies: Real-World Examples of Successful AI Integration

The following case studies in healthcare, finance, and retail highlight how AI enhances human capabilities and drives better outcomes.

Case Study 1: Healthcare and AI Diagnostics

In the healthcare sector, AI is making significant strides in transforming medical diagnostics, particularly through advanced image analysis. A prime example is RETFound, an AI foundation model developed by researchers at UCLA and Moorfields Eye Hospital that represents a major leap forward as one of the first AI foundation models in healthcare. Trained on a dataset of 1.6 million retinal scans at Moorfields Eye Hospital, RETFound utilizes AI tools and infrastructure provided by the NHS-led health data research hub for eye health. The model identifies over 50 sight-threatening eye diseases, including diabetic retinopathy and glaucoma, and predicts systemic health conditions, such as heart attacks, strokes, and Parkinson's disease. RETFound outperforms existing AI systems across a range of complex clinical tasks while maintaining effectiveness in diverse populations—an area where many AI models have traditionally struggled.

One of RETFound's most innovative aspects is its self-supervised learning approach, which dramatically reduces the need for extensive human labeling. By predicting missing portions of images with minimal expert input, RETFound achieves comparable performance with just 10% of the usual labeled data. Moreover, by making RETFound available as an open-source tool, the research team is accelerating global efforts to prevent blindness and improve health outcomes, garnering international opportunities to leverage RETFound for new studies.

Case Study 2: Finance and Revolutionizing Financial Operations with AI

In finance, where efficiency and accuracy are paramount, AI is indispensable for managing large-scale data analysis, detecting fraud, and automating routine operations. One of the most notable advancements is JPMorgan Chase's AI-powered tool, COiN (contract intelligence), which has revolutionized the review of legal documents. It can process 12,000 commercial credit agreements in seconds—a task requiring 360,000 hours of manual labor (JPMorgan Chase, 2017). This tool exemplifies how AI can streamline complex processes, freeing financial professionals to focus on more strategic activities.

AI has also transformed customer interactions in the finance industry by personalizing experiences and increasing engagement through technology. AI-powered virtual assistants and chatbots provide 24/7 customer service and recognize signs of potential fraud. These tools respond to customer queries in real-time, enhancing operational efficiency. The rise of AI-driven personalization in banking enables institutions to offer tailored financial advice and product suggestions based on individual financial activities and goals, thus revolutionizing how financial institutions serve their customers.

Case Study 3: Retail and Customer Experience

In the retail sector, AI has revolutionized the customer experience by enabling personalized shopping and support, allowing retailers to meet the demands of digitally savvy consumers. Sephora, a global beauty retailer, has integrated AI solutions into its customer service strategy through an AI-powered tool that allows customers to try on makeup virtually using facial recognition technology and then offer personalized product recommendations based on past purchases. Amazon's AI-powered recommendation engine sets the standard in the retail industry by analyzing customer data, including past purchases and browsing behavior, to deliver tailored product suggestions that boost conversion rates.

AI's role optimizes core retail operations, including inventory management, demand forecasting, route planning, price optimization, and assortment planning. AI-driven inventory management systems combine customer purchase data with supply chain analytics to predict future buying trends, align stock levels, and eliminate inefficiencies, reducing waste and improving profitability. Retailers increasingly invest in AI to stay competitive. By refining operations and engagement models through AI, retailers position themselves to thrive in a rapidly evolving, digital-centric commerce environment.

These case studies illustrate the transformative potential of AI across diverse sectors, highlighting how AI enhances human capabilities and drives superior outcomes when intentionally integrated. AI can be a catalyst for innovation, driving meaningful improvements in business operations and customer engagement. These case studies serve as compelling evidence of AI's potential to transform industries and redefine the future of work.

Addressing Challenges and Ethical Considerations

As AI continues to permeate the workplace, it is crucial to recognize its potential and limitations. While AI offers the promise of transforming industries and improving efficiency, it also poses significant ethical and operational challenges. Addressing these challenges is vital to ensuring AI benefits everyone and does not exacerbate existing inequalities.

Recognizing AI's Limitations

Despite their advanced capabilities, AI systems are not infallible. A major concern is the potential for bias in decision-making processes. AI models are often trained on historical data, which can contain ingrained biases related to race, gender, or socioeconomic status. These biases can be inadvertently perpetuated by AI, leading to discriminatory outcomes in critical areas such as hiring, promotions, and performance evaluations.

Another significant challenge is data privacy. AI systems rely on vast amounts of data to function effectively, raising concerns about how personal and sensitive information is collected, stored, and used. The improper handling of such data can lead to breaches and misuse, compromising the privacy of employees and customers alike. Ensuring robust data protection is essential to maintaining trust and safeguarding individual rights.

Building a Fair and Inclusive AI Environment

To mitigate the risks associated with AI, it is imperative to create a fair and inclusive AI environment. One of the most effective strategies is to ensure diverse representation within AI development teams. When AI development includes a wide range of perspectives—spanning different genders, ethnicities, and socioeconomic backgrounds—the likelihood of producing fair and unbiased algorithms increases.

In addition to diverse teams, the use of datasets that accurately represent the population is crucial in reducing bias. Algorithms trained on representative data are less likely to produce skewed outcomes that disadvantage certain groups. Furthermore, transparency in AI systems is essential. When the decision-making processes of AI are open, it allows for greater scrutiny by stakeholders, ensuring AI decisions are just and equitable.

Final Thoughts

Integrating AI in the workplace is not just a technological shift—it is a profound transformation that challenges traditional concepts of work, productivity, and human potential. As AI continues to advance, it is essential to approach this transition with both optimism and caution. While AI offers unparalleled opportunities to enhance efficiency, creativity, decision-making, and human relationships, it also presents significant ethical and practical challenges. The key to successfully navigating this new landscape lies in balancing innovation with responsibility, ensuring that AI amplifies rather than diminishes human capabilities.

As we stand on the brink of an AI-driven future, the path forward is clear: embrace AI not as a replacement, but as an enhancement of human potential. The successful integration of AI into the workplace requires a multifaceted approach that involves recognizing AI's limitations, building inclusive and fair AI systems, adhering to ethical guidelines, and promoting responsible AI use. Responsible use also creates opportunities for intentionally building productive and trusting relationships with clients, patients, and customers. By doing so, organizations can ensure that AI is a tool that empowers employees, drives innovation, and creates a more equitable workplace. As AI continues to evolve, so too must our approach to work—embracing change, advocating for fairness, and ensuring AI benefits are shared by all.

References

Binns, R. (2023). Bias in AI: Understanding and mitigating discriminatory practices in automated decision-making. *Journal of Ethics and Technology, 15*(2), 102–118.

Chui, M., Hazan, E., Roberts, R., Singla, A., Smaje, K., Sukharevsky, A., ... Zemmel, R. (2023). The economic potential of generative AI: The next productivity frontier. *McKinsey & Company*. Retrieved from https://www.mckinsey.com/capabilities/mckinsey-digital/our-insights/the-economic-potential-of-generative-ai-the-next-productivity-frontier#key-insights

De Fauw, J., Ledsam, J. R., Romera-Paredes, B., et al. (2018). Clinically applicable deep learning for diagnosis and referral in retinal disease. *Nature Medicine, 24*, 1342–1350. https://doi.org/10.1038/s41591-018-0107-6

Downey, S. (2024). The godmother of AI: How Fei-Fei Li '99 is safeguarding the future of human and artificial intelligence. *Princeton Alumni Weekly*. Retrieved from https://alumni.princeton.edu/stories/fei-fei-li-woodrow-wilson-award

Dweck, C. S. (2006). *Mindset: The new psychology of success.* Random House Publishing Group.

Floridi, L., & Cowls, J. (2022). The ethics of AI: An overview of the foundational principles and practical guidelines. *Ethics and Information Technology, 24*(3), 247–263.

Jobin, A. (2023). Fostering a culture of ethical AI use: The role of ethics committees and employee training. *Journal of Business Ethics, 169*(2), 385–402.

Khan, S. A. (2024). Sundar Pichai's top leadership principles revealed. *YourStory*. Retrieved from https://yourstory.com/2024/06/success-blueprint-sundar-pichai

Li, T. (2024). AI in fintech: The role of artificial intelligence in transforming the finance industry. *Devot Team*. Retrieved from https://devot.team/blog/ai-in-fintech

Microsoft News Center. (2023). Moody's and Microsoft develop enhanced risk, data, analytics, research and collaboration solutions powered by generative AI—Stories. *Microsoft News*. Retrieved from https://news.microsoft.com/2023/06/29/moodys-and-microsoft-develop-enhanced-risk-data-analytics-research-and-collaboration-solutions-powered-by-generative-ai

Son, H. (2017). JPMorgan software does in seconds what took lawyers 360,000 hours. *The Independent*. Retrieved from https://www.independent.co.uk/news/business/news/jp-morgan-software-lawyers-coin-contract-intelligence-parsing-financial-deals-seconds-legal-working-hours-360000-a7603256.html

UCL News. (2023). World-first AI foundation model for eye care to supercharge global efforts to prevent blindness. *University

College London. Retrieved from https://www.ucl.ac.uk/news/2023/sep/world-first-ai-foundation-model-eye-care-supercharge-global-efforts-prevent-blindness

Bitter, A., Luna, N., Mayer, G., Dawkins, J., & Reuter, D. (2023). 6 retailer leaders innovating the customer experience with AI. *Business Insider*. Retrieved from https://www.businessinsider.com/people-revolutionizing-customer-experience-ai-2023-10

Spencer, A. (2024). Artificial intelligence in retail: 6 use cases and examples. *Forbes*. Retrieved from https://www.forbes.com/sites/sap/2024/04/19/artificial-intelligence-in-retail-6-use-cases-and-examples/

West, S. M. (2023). The importance of diversity in AI development: A path to fair and equitable systems. *AI and Society, 38*(1), 45–61.

About the Author

Dr. Emily Barnes is an expert in artificial intelligence and a distinguished academic researcher, educator, and adminstrator with over 15 years of experience in technology, AI, and higher education administration. As one of the few prominent female voices in the intersection of AI and education, she has held pivotal roles, including chief digital learning officer, provost, and president, where she has championed the transformative power of education and student-centered learning. Dr. Barnes holds a PhD in Artificial Intelligence from Capitol Technology University and an EdD in Higher Education Administration. She is the creator of the AI8-Model, a groundbreaking framework for change management, assessment, and implementation across organizations. Her work focuses on leveraging AI to enhance learning environments, advocating for women in STEM, and promoting inclusivity in the tech industry. A sought-after speaker and consultant, Dr. Barnes is dedicated to driving innovation and shaping the future of AI in education.

Connect with Dr. Emily J. Barnes via LinkedIn at https://www.linkedin.com/in/emilyjbarnes/ or visit her website at emilyjbarnes.com. Research portfolio at https://www.researchgate.net/profile/Emily-Barnes-30, https://orcid.org/0000-0001-9401-0186. For inquiries, she can be reached at dremilyjbarnes@gmail.com.

CHAPTER 4

FROM VISION TO IMPACTFUL AI INNOVATION

By Hans van den Berg
Managing Partner at SAY HAI
Breda, The Netherlands

Great things are done by a series of small things brought together.
—Vincent van Gogh

The Importance of a Structured AI Framework

Today, in a business landscape defined by rapid change, AI isn't just a "nice to have"—it's an essential catalyst for transformation. Yet, many organizations make a fundamental mistake: they dive into AI projects without first understanding their core purpose. They launch scattered, isolated initiatives, believing technology alone will solve their problems. But without a clear "why," these efforts often fall short of their true potential. The real question isn't just what AI can do, but why we need it and how it can drive purpose-led change.

What's missing is a strategic approach that begins with an organization's mission, vision, and values. Implementing AI shouldn't be about chasing trends or keeping up with competitors; it's about ensuring every AI initiative is deeply rooted in what your business stands for. This is where a structured framework becomes indispensable.

This framework acts as a guiding compass, ensuring AI isn't just another tech endeavor but a strategic enabler of long-term value. It helps organizations navigate technological shifts with clarity and intent. By understanding your "why," the framework allows you to adapt and find the best path forward while staying aligned with your ultimate goals.

In this chapter, we'll walk through a five-phase process designed to help you adopt AI in a way that's meaningful, impactful, and sustainable. We'll explore discovery, design, implementation, deployment, and measurement—all within the context of an AI-driven world where agility and adaptability are essential for future success.

Introducing the Impactful AI Innovation Framework

The Impactful AI Innovation Framework is not just a roadmap—it's a mindset. A mindset that focuses not only on leveraging AI but on ensuring that AI becomes a powerful enabler for achieving your organization's bigger purpose. It's about creating an environment where AI isn't siloed into different departments or isolated projects, but integrated into the DNA of your company. AI, when harnessed properly, has the potential to bridge the gap between what your business is today and what it could become tomorrow.

The framework provides a structured approach to ensure that every step of your AI journey is aligned with your strategic objectives. It ensures that AI initiatives are not only innovative but also meaningful—delivering real value, transforming operations, and enhancing the human experience. The process consists of five key phases:

1. *Discovery*: understanding where you are and where you can go

2. *Design*: creating an AI compass and actionable AI game plan that aligns with your vision

3. *Implement*: turning your vision into reality starting and testing and finding the true business impact

4. *Deploy*: scaling AI initiatives to drive tangible results

5. *Measure*: continuously assessing and improving your AI initiatives to ensure they deliver on your impact goals

This framework gives leaders the confidence that every AI investment is tied to something bigger—whether it's enhancing customer experience, improving operational efficiency, or driving innovation.

Phase 1: Discovery

The first step in any journey is knowing where you are. In the discovery phase, we focus on gaining a comprehensive understanding of your current landscape. This means looking at your operating model, which is the backbone of how your organization runs, or focusing on the customer journey, where you engage and interact with your stakeholders, the customers and or consumers.

But the discovery process goes deeper than just identifying processes and stakeholders. It's about taking a high-level look at your business through the lens of the AI compass. This involves assessing key components like ethics, sustainability, and alignment with your broader business objectives. What are the current challenges you're facing? Where are the opportunities for growth and transformation? Who will be affected by the changes AI will bring, both internally and externally?

In this phase, we're not yet diving into specific solutions or AI technology. We're setting the stage. We're ensuring that when we do move forward, we're solving the right problems with the right tools. Discovery is about framing the challenge and building a shared understanding across the organization of what needs to be achieved and why.

This phase sets the foundation for everything that follows, ensuring that we start with a clear purpose and direction.

Phase 2: Design

In the design phase, we design the AI compass, which includes several core components that guide the organization's AI journey. These components ensure that every AI initiative aligns with the company's mission, vision, and values.

AI Compass Components:

1. **Ethical AI**: this ensures that AI initiatives are fair, transparent, and responsible. Ethics guide decision-making processes and help avoid biases in AI applications. Regulatory Compliance

ensures that all AI related processes, systems, and technologies within an organization adhere to relevant laws, industry standards, and ethical guidelines.

2. **Sustainability**: AI initiatives must be environmentally conscious, minimizing energy consumption and promoting sustainable practices.

3. **Human-Centric AI**: AI should enhance human capabilities, not replace them. It should be used to support people in achieving greater efficiency and impact.

4. **Business Alignment**: AI projects must align with long-term business goals, ensuring that AI is directly tied to driving meaningful business outcomes.

5. **AI Game Plan**: this is where we identify specific AI initiatives, mapping out how AI will solve key challenges and create value for the organization on top we objectively prioritize the initiatives.

Business Impact Canvas

Within the AI game plan design we use the business impact canvas to assess and document the pains and gains for the organization. By evaluating these factors, we can identify where AI can have the most significant positive impact on the business. The canvas breaks down key areas such as current inefficiencies (pains) and opportunities for innovation (gains), ensuring that AI projects target the right areas for improvement.

Here's how to use it:

1. Identify and describe the pains and gains by determining the problems or wishes of particular stakeholders and why.

2. Document separately the root causes to understand underlying problems.

3. Assess the impact of the pain or gain in terms of cost, time, and process delays.

4. Define the nirvana situation and, thus, the desired outcome. Define the ideal state if the pain is resolved or the gain is realized.

This structured approach ensures that AI initiatives are aligned with business objectives and stakeholder needs, facilitating effective and strategic AI integration.

Once AI initiatives are identified and documented, we use two key methods to prioritize them. The AI Compas can contain more company prioritization methods.

1. **Impact and Complexity Matrix:** this matrix assesses each initiative based on its potential business impact and the complexity of implementation. High-impact, low-complexity initiatives are prioritized, ensuring that the organization gets the most value early on while managing the level of risk.

2. **Innovation Focus Model:** this model helps determine whether initiatives are internally or externally focused and whether they are solving current problems or creating future opportunities. This ensures that the organization has a balanced approach to both improving current processes and driving innovation for the future.

In the design phase, the AI compass and these tools guide us in creating a robust, actionable roadmap that aligns with the organization's strategic direction while driving measurable, impactful results.

Phase 3: Implement

Ideas alone don't drive change—implementation does. But implementation isn't about diving into large-scale integration immediately. Instead of a traditional proof of concept (POC), we focus on proof of impacts. It's about taking small, calculated steps to test the waters without burning resources, people's time, or money. This phase is designed to measure the exact impact of AI initiatives on a small, controlled group before committing to a larger rollout.

The key here is to execute in a secure, limited environment—where the AI isn't yet integrated into the full ecosystem. It's about setting up AI with a select group of users, often internally, where we can carefully monitor outcomes, make adjustments, and validate the true business impact. This approach keeps risk low while allowing for learning and iteration.

In this phase, the goal is not to disrupt but to enable. We're preparing your organization, processes, and people to embrace AI in a way that feels natural, impactful, and non-invasive. By working with key teams, we ensure they have the skills and resources needed to incorporate AI into their workflows when it's ready for full deployment.

In this small-scale environment, we test the assumptions made in the design phase. Are we seeing the expected results? Does AI deliver the impact we projected? Are there unexpected hurdles? This proof of impact approach allows us to validate and refine AI in real-time, ensuring that when full deployment arrives, it's not just functional but transformational.

It's about creating real value step by step, with every move measured, purposeful, and aligned with the broader organizational goals.

This is where the vision starts to become reality and where the organization begins to see the tangible benefits of AI.

Phase 4: Deploy

With the AI initiatives tested and refined, it's time to deploy them across the organization. But we don't do this all at once. The deploy phase is about rolling out AI in a controlled, strategic manner—starting with a small group of users or internal stakeholders. By doing this, we ensure that any remaining challenges are addressed before scaling AI to a broader audience.

Deployment is not just about flipping a switch and letting AI run. It's about ensuring that the technology, people, and processes are all aligned to support the AI initiatives effectively. This phase involves gathering feedback, adjusting the AI Technology based on real-world use, and ensuring that the deployment is seamless and efficient.

It's also about proving the value of AI in a manageable, low-risk way before expanding across the organization. This phased approach allows for continuous learning and improvement, ensuring that AI not only delivers on its promise but does so in a way that's sustainable and scalable.

The deploy phase is where AI starts to make a real impact, but it's also where we make sure everything is working as expected before scaling up.

Phase 5: Measure

The final phase is measure, but it's by no means the end. In fact, this phase is what keeps your AI initiatives dynamic, adaptive, and continuously improving. Measuring impact isn't a one-time activity—it's an ongoing process that ensures AI is delivering value and remains aligned with the organization's evolving goals.

In this phase, we evaluate the impact of AI initiatives against what was defined during the design phase. Did AI achieve the desired business impact? Is it still aligned with the ethical standards and sustainability goals we set? Are we in sync with our AI Compass?

This phase ensures that AI is achieving the expected results and remains adaptable to future challenges. By continuously measuring, we keep the AI initiatives aligned with the evolving business goals and strategic direction of the organization. But it's not just about measuring success—measure is about learning. It's about iterating on what works, adjusting what doesn't, and ensuring that AI initiatives continue to drive real impact.

This phase brings the AI journey full circle. It's about learning, growing, and ensuring that AI continues to serve the organization in meaningful and impactful ways. AI is never static—it's always evolving, just like your business. And that's why measure isn't the end of the process; it's a constant feedback loop that keeps your organization at the forefront of impactful AI innovation.

Values

Without a clear set of guiding principles, AI can easily become a tool for short-term gains with long-term consequences. In this framework, the values are what tether AI to the organization's higher purpose, ensuring that every initiative aligns with what truly matters. At the heart of our framework, we follow these values:

1. **Less Is More**: simplicity drives focus. In the world of AI, complexity can often become the enemy of progress. We believe in the power of simplicity—focusing on solutions that are lean, effective, and purposeful. When AI initiatives are simple, they are scalable, adaptable, and far more impactful.

2. **Energy and Planet Aware**: every AI decision has an environmental impact. As stewards of our planet, we must ensure that AI technologies are sustainable, energy-efficient, and mindful of their ecological footprint. This value reminds us that true innovation doesn't come at the cost of our environment.

3. **Responsible AI**: AI should always be human-centered. Responsible AI isn't just a buzzword—it's a necessity. From ensuring fairness and transparency to safeguarding against bias, this value ensures that AI enhances human life and supports ethical decision-making at every step. We believe in IA: intelligent augmentation where the technology supports humans

4. **Business Impact Driven**: AI must drive tangible business results. At the end of the day, every AI initiative must deliver measurable impact. Whether it's improving customer experience, reducing costs, or driving operational efficiency, this value ensures that AI isn't just a tech solution—it's a business strategy.

These values are the guiding compass that ensures AI isn't just about technology—it's about purpose. They provide the ethical framework needed to ensure that AI initiatives deliver real impact while staying true to the organization's mission.

Use Cases | Domains

AI is transforming industries across the board. But what truly matters isn't the technology itself—it's how AI is applied to solve real-world problems and create new opportunities. Let's explore how AI is making an impact across different domains, providing clear examples of what's possible when AI is aligned with business strategy.

1. **Machine Learning**: predictive analytics is at the core of machine learning, enabling businesses to anticipate trends, optimize operations, and make smarter decisions. In healthcare, for example, machine learning models are being used to predict patient outcomes, personalize treatment plans, and streamline administrative processes. Retailers are using machine learning to forecast demand and optimize inventory, reducing costs and improving customer satisfaction. The impact of machine learning isn't just about data—it's about turning data into actionable insights that drive the business forward.

2. **Machine Vision (Computer Vision)**: machine vision technologies are revolutionizing industries like manufacturing and healthcare by automating quality control and improving diagnostic accuracy. In factories, AI-powered cameras can detect defects

in products in real time, ensuring that only the highest-quality goods reach the consumer. In healthcare, machine vision is being used to analyze medical images like X-rays and MRIs, helping doctors diagnose diseases more accurately and faster than ever before. The power of machine vision lies in its ability to see what humans can't—and turn that vision into action.

3. **Conversational AI**: the way we interact with machines is changing, and conversational AI is at the forefront of this shift. From customer service assistants to virtual assistants, AI-powered conversations are helping businesses streamline communication and improve customer experience. In the financial industry, for example, banks are using AI chatbots to handle routine customer inquiries, freeing up human agents to focus on more complex tasks. By automating repetitive tasks, conversational AI allows businesses to scale their operations while still maintaining a personal touch.

4. **Generative AI**: the creative potential of AI is perhaps most evident in generative AI, which can produce everything from text to images, videos, and music. In marketing, generative AI is being used to create product content like inspirational stories and SEO content. Creating personalized ads and social media content at scale helps brands engage with their audiences in new and meaningful ways. The possibilities with generative AI are endless, but the key is in how businesses harness its creative power to connect with their customers and consumers.

These use cases demonstrate the incredible versatility and potential of AI. But more importantly, they show how AI, when aligned with business strategy, can drive real-world impact. Whether it's improving operational efficiency, enhancing customer experience, or unlocking new creative opportunities, AI has the power to transform industries and change the way we work and live.

Conclusion

The AI journey is one of constant discovery, design, implementation, and refinement. But it's not just about the technology—it's about aligning AI with your organization's purpose, values, and long-term strategy. The Impactful AI Innovation Framework provides the structure and the guidance needed to navigate this complex landscape, ensuring that AI initiatives not only deliver tangible results but also serve a greater mission.

As the business world continues to evolve, the organizations that thrive will be the ones that approach AI with purpose, vision, and a deep commitment to driving real impact. The future is not about doing more with AI—it's about doing AI right.

About the Author

With a deep passion for education, Hans van den Berg believes in the power of learning to drive change, aligning with Nelson Mandela's words, "Education is the most powerful weapon to change the world." Starting with a background in teaching electrical engineering, Hans has built a global career, managing educational businesses, and later co-founding a B2B SaaS company focused on product information and project portfolio management. Currently, he leads the Data and AI agency of the Handpicked Agency Group, a collective of 13 specialized agencies in the Netherlands. Driven by his belief that AI should support, not replace, people, Hans emphasizes small, impactful steps in AI adoption, ensuring that technology aligns with real business needs. With experience across multinationals and industries like retail and manufacturing, his mission is to guide businesses through ethical and effective AI adoption.

Email: hans@sayhaito.ai
Website: www.sayhaito.ai
LinkedIn: https://www.linkedin.com/in/hansvandenberg/
Find out how you personally can innovate with AI on www.sayhaito.ai/aimindset

CHAPTER 5

THE COGNITIVE ADAPTATION

By Monika Bishnoi
AI Product Manager
San Francisco, California

> *Intelligence is the ability to adapt to change.*
> —Stephen Hawking

The Adoption Gap

Working for SAP, the world's largest enterprise software company, I've witnessed how software plays a critical role in managing business processes ranging from finance to human resources, from procurement to supply chains, and from manufacturing to customer relations. In a sense, these software systems are as vital to a company as the central nervous system is to our bodies. To illustrate the impressive scale of SAP's impact, these applications touch over 87% of the world's global commerce. In other words, the odds are that nine out of ten products you know or use, at some point in their journey from manufacturing to consumption, have been touched by one of these applications.

Over the past year, I've witnessed a fascinating and humbling phenomenon unlike anything I've seen in my decade of experience at SAP. As an AI product manager, I saw nearly *all* of these applications "extended" with the latest advancements in AI. This dramatic rise of AI applications led me to contemplate how to make sense of it all. Was there an underlying pattern to understand these diverse "applications" of AI? Or, better yet, could I generalize the "applicability" of AI to facilitate better adoption of this technology?

Beyond my personal quest, I also wanted to explore a more pressing question: is AI's hype truly real? Sequoia Capital's article "AI's $600B Question" highlights the growing gap between AI infrastructure expenditure and revenue (exceeding $500 billion as of this writing), questioning the sustainability of AI's boom. Other institutions like Goldman Sachs Research Newsletter ask, "GenAI: Too much spend, too little benefit?" concluding that adoption and product-market fit will take longer than anticipated. Regarding the latest generative AI wave, skeptics argue that while it's impressive to see AI models generate Beatles-esque songs, innovative recipes that actually taste good, and videos that could make even Hollywood blush, how will these abilities truly impact the productive functioning of the world?

In search of answers, I began developing a comprehensive framework to help myself and others grasp AI's vast potential. If we truly want to develop an AI Mindset, we need to go from mere adoption of this technology to *adapting* our thinking to this coming wave of Artificial intelligence, what I call *cognitive adaptation*. This chapter offers an insider's perspective on how to adapt to AI, exploring how it's revolutionizing our abilities today and shaping our tomorrow.

Let's begin.

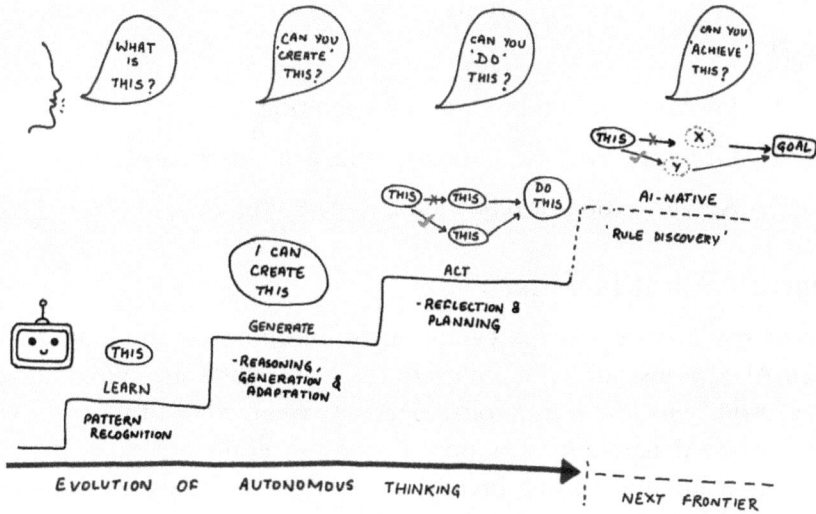

Illustration by Monika Bishnoi

Evolution of Autonomous Thinking

To understand how this technology extends our capabilities, let's first define it. Here's my "tl;dr" version: AI, in its current form, is a general-purpose technology. Just as electricity (another general-purpose technology) powers the autonomous movement of all kinds of physical systems using motors, AI powers "autonomous thinking" in all kinds of information systems using models.

While AI has existed for some time, it hasn't always been general-purpose. In *The Vital Question*, biochemist Nick Lane argues that "energy constrains the evolution of life on Earth." He further suggests that the interplay of energy and evolution could form the basis of predicting biology. Applying this analogy to AI systems, we see how the energy behind AI—namely, the availability of data, compute power, and underlying model architectures—constrained its evolution into a general-purpose technology. Exploring the history of these constraints allows us to understand the evolution and predict the coming wave of AI applications.

Generally, three evolutionary steps in autonomous thinking have emerged:

1. **Learn**: evolution of pattern recognition
2. **Generate**: evolution of reasoning, generation, and adaptation
3. **Act**: evolution of reflection and planning

Learn: "What Is This?"

The first major step in the evolution of autonomous thinking arrived when AI demonstrated the ability to recognize patterns. While humans excel at this, we have limitations, especially when confronted with large, complex, or dynamic information. For instance, most people can identify only a handful of dog breeds and struggle to recognize hundreds. Given sufficient data, a simple program, on the other hand, can identify the underlying complex patterns of dog breeds and recognize nearly all with remarkable precision. This is called machine learning. Like a child learning to identify objects through repeated exposure, machine learning models seem to grasp the underlying patterns from a set of examples (or labeled data). The model appears to *learn*. This means the programs may learn patterns that are known to us, but also patterns that are new to us.

This capability represents a significant advancement in autonomous thinking. Any task involving *identification, prediction, classification,* or *matching* can benefit. This is the same powerful capability that enables self-driving cars to identify objects on the road and smartphones to recognize faces for unlocking.

In recent years, AI has also been applied to various business tasks, such as matching people to jobs, products to customers, teams to appointment slots, and ad prices to web page placements. It's also used to predict customer/employee churn, warehouse inventory, and stock prices, or to identify complex information in visual inspections of machines or medical images. We can now extend our ability to understand things at scale and speed like never before. The next time you find yourself asking, "What is this?", "What are the patterns in my finances?", or "Can I stock products better by looking at patterns of my annual sales?" consider AI; you might be surprised by the hidden patterns it reveals.

The ability to learn specific patterns for a given data is also, however, its limitation. A model that recognizes hundreds of breeds of dogs may fail to identify patterns in molecular structures for vaccine development.

As these *specialized* models are narrow in their applicability (also called "narrow AI"), they require significant time and effort to be used effectively. This is where the next AI breakthrough emerges, representing the next step in the evolution of autonomous thinking.

Generate: "Can You Create This?"

The evolution to "generate" stems directly from scaling the underlying models, data, and computation. While scaling has existed in fields like high-performance computing and neuroscience, its specific application to AI gained prominence in the late 2010s. Proponents of the "scaling hypothesis" demonstrated its success with large-scale language models like OpenAI's GPT-3 in 2020.

The "generate" step showcases AI's creative potential by producing novel content based on learned patterns. Large language models, which have been the focus of much attention, predict the next most probable word or phrase, arranging words to convey meaning without rigid rules. This same predictive model generates novel outputs (to us) when applied to other modalities like images and sound.

Interestingly, these models demonstrate emergent capabilities of reasoning and adaptation, mirroring perhaps how humans use language to objectify, understand the world, reason, and adapt. By making sense of our language, AI models seem to have unlocked something deeper. As Ilya Sutskever, chief scientist and co-founder of OpenAI, notes, "AI has learned a world model." Like a child progressing from object identification to language acquisition, sentence generation, thought expression, reasoning, and adaptation to instructions, these AI models demonstrate a remarkable leap in autonomous thinking. We now have general-purpose "virtual brains" capable of *understanding human language, reasoning, generating, adapting,* and becoming our ultimate interface to any information system.

Models like those powering ChatGPT are already creating immense value in complex business applications, such as generating new content (software code, marketing copy, translations, images), expanding existing information (product/job descriptions, contracts, proposals, emails), summarizing (research, industry news, customer profiles), and engaging in question-and-answer sessions (financials, product manuals, HR policies, or business activities).

These models also have limitations. As word prediction machines trained on vast and sometimes messy data, they can provide factually incorrect answers ("hallucinations") and inherit human biases. They sometimes struggle with simple questions, like incorrectly stating 9.11 is larger than 9.9 (amusingly referred to as "jagged intelligence" by Andrej Karpathy). Furthermore, these models produce "one-shot" responses, which Andrew Ng likens to writing about a topic without rewriting or revisiting. Researchers are exploring methods like prompting and grounding to verifiable information sources to improve the response quality of these models. One solution that significantly enhances output is providing the model with a "guide" or "recipe" (also called chain-of-thought prompting or reasoning) specifying the desired output. This leads to the emergence of a new kind of AI application, which is also the next step in the evolution of autonomous thinking.

Act: "Can You Do This?"

The next evolution in autonomous thinking is the shift from passive generation to active engagement. With the evolution to act or "ReAct" as described in the foundational paper, AI can act autonomously with a certain degree of "agency," resulting in what are known as AI agents. AI agents enhance existing AI models by enabling them to *reflect* (self-monitor actions and evaluate responses) and *plan* (iteratively execute instructions using available tools). Similar to a child developing reasoning and improving skills like drawing or playing piano, an AI agent, given an objective, can not only understand it but also make choices, assess actions, and adapt behavior based on feedback.

This stage represents a significant leap in AI capabilities. Companies like Adept AI and AutoGPT are developing general-purpose AI agents for enterprise productivity and task completion. Others are exploring domain-specific AI assistants, such as Harvey for legal assistance and Abridge for medical records and doctor-patient conversations. In business, AI agents can fulfill various assistant roles, such as an executive assistant (managing time, optimizing calendars, writing emails), a production and facility assistant (optimizing machine functioning hours), or a customer service assistant (planning, scheduling, and responding to customer inquiries).

While still in their infancy, AI agents hold immense promise. So, if you find yourself asking, "How can I *do* this?", "How can I source the best materials?", or "How can I plan my trip?"—essentially seeking ways to accomplish specific objectives—AI agents might be the allies you've been looking for.

THE NEXT FRONTIER

AI-Native: "Can You *Achieve* This?"

We've explored the evolutionary steps in autonomous thinking that dominate current AI applications. However, these applications are primarily "extensions" of existing workflows. The true power of AI lies not in extending existing applications but in creating entirely new workflows—what I call *AI-native* applications, the next, largely unexplored frontier in autonomous thinking.

This insight stems from a deep understanding of AI's power, specifically the limitations of explicit optimization and the power of implicit goal-oriented learning. It means a fundamental shift from a linear, static process view to a dynamic, adaptive one that recognizes the unique path each of us takes towards achieving a goal.

AI-native applications leverage AI's full potential. Unlike AI agents, which are defined by instructions and available tools, AI-native applications, built explicitly for a goal, *identify rules* (instructions connected by cause and effect) to achieve our goals. These systems evolve from mere pattern discovery to *rule discovery*.

This isn't an entirely new concept. The $600 billion digital advertising market has traditionally relied on rule-based systems for ad placement. However, dominant players like Google and Meta have built AI-driven systems that allow marketers to set revenue or profit goals and let AI handle the rest. The AI identifies the best audience, creates personalized creatives, and selects the optimal channel for conversion. This AI-native approach is redefining the traditional marketing funnel (perhaps eliminating it altogether), and customers using these systems are already reaping significant benefits.

Another successful AI-native system is found in autonomous vehicles, where the complexity of dynamic variables like road and traffic conditions makes explicitly coding the driving rules from point A to point B impossible. The desired behavior of these applications is learned from training data and "encoded" in the AI model itself, a concept referred to as "software 2.0" by Andrej Karpathy.

Built for a Goal, Not the Process

The future of AI-native business software is unfolding, but with AI at the heart of company workflows, results could be achieved with remarkable efficiency. AI could navigate complex and dynamic process details, choose the optimal path through rule discovery, execute tasks using tools, keep us informed of decisions, and ultimately help us achieve our goals more effectively. This AI-native program, which I call a "Unit," would understand our business goals and autonomously plan to achieve them.

Imagine an AI-native CRM application called "RevGen" Unit, designed to optimize revenue generation, capable of tapping into the non-linear, dynamic, and complex tasks faced by salespeople. Sales teams could define their goals and let RevGen handle the rest. They could focus on building meaningful customer relationships while RevGen suggests next steps, prepares for meetings, books travel, creates quotes, sends emails, and tracks progress towards revenue targets. Now imagine similar Units for procurement, manufacturing, marketing, HR, etc., each interacting and optimizing for the ultimate company goal: sustainable profits. Imagine these Units also "evolve" with the ever-evolving needs of the business. This could eliminate systemic redundancies and inefficiencies across all kinds of organizations, big or small, creating a trillion-dollar global impact and a multi-billion-dollar opportunity for AI-native disruptors to revolutionize how the world works.

The Cognitive Adaptation

Now that we've explored the current and the future opportunities of AI through the evolution of autonomous thinking, the next logical question is: what will be our role in relation to this evolution? Does this replace the need for humans for all kinds of cognitive work? Experts are divided into optimists—those who see the glass half full—and pessimists—those who see the glass half empty.

A realist's perspective requires flipping the lens to consider the evolution of human intelligence (and not AI), wielding autonomous thinking as a tool. The question then becomes: if our intelligence can create artificial intelligence, can it also adapt to it? If so, will we be well adapted, leading to a utopia, or maladapted, leading to a dystopia? Regardless, one thing we know for sure is we *will* adapt our thinking to AI to either become AI supervisors or be supervised by AI.

Charles Darwin once said, "It is not the strongest of the species that survives, nor the most intelligent. It is the one that is most adaptable to change." With this fast-approaching change, what's important is to understand what it is and where we can leverage this autonomous thinking to augment our intelligence and make a positive step towards the coming cognitive adaptation.

References

World Economic Forum. SAP. https://www.weforum.org/organizations/sap-se/

David Cahn. (2024). AI's $600B Question. Sequoia Capital. https://www.sequoiacap.com/article/ais-600b-question/

Allison Nathan, et al. (2024). *GEN AI: TOO MUCH SPEND, TOO LITTLE BENEFIT?*. The Goldman Sachs Group, Inc. https://www.goldmansachs.com/images/migrated/insights/pages/gs-research/gen-ai--too-much-spend%2C-too-little-benefit-/TOM_AI%20 2.0_ForRedaction.pdf

Nick Lane. (2015). *The Vital Question: Energy, Evolution, and the Origins of Complex Life*. W. W. Norton & Company

Jensen Huang, Founder and Chief Executive Officer, NVIDIA & Ilya Sutskever, Co-founder and Chief Scientist, OpenAI (2023). *Fireside Chat with Ilya Sutskever and Jensen Huang: AI Today and Vision of the Future*. https://www.nvidia.com/en-us/on-demand/session/gtcspring23-s52092/?ncid=so-yout-561702

Andrej Karpathy. (2024). *Jagged Intelligence*. https://x.com/karpathy/status/1816531576228053133?lang=en

Shunyu Yao, et al. (2023). *ReAct: SYNERGIZING REASONING AND ACTING IN LANGUAGE MODELS*. Department of Computer Science, Princeton University 2 Google Research, Brain team. https://arxiv.org/pdf/2210.03629

Andrej Karpathy. (2017). *Software 2.0*. https://karpathy.medium.com/software-2-0-a64152b37c35

About the Author

Monika Bishnoi is a seasoned AI product manager with over a decade of experience at SAP, one of the world's largest enterprise software companies. She spearheads AI product innovations, strategy, and road-map across SAP's entire product portfolio, encompassing hundreds of applications for diverse business processes and industries. Monika's rare blend of technical expertise and business acumen stems from her computer science engineering degree and MBA from the prestigious Indian Institute of Management. Her work with hundreds of companies across continents and industries has earned her numerous accolades, including recognition as SAP's Future Leaders and awards for innovation in product management. Monika's passion for AI, which began during her undergraduate studies, continues to drive her professional endeavors. As the founder of Docents, an educational platform on YouTube with over 250,000 views, she demonstrates her commitment to sharing knowledge and fostering understanding of complex concepts.

Email: contact@monikabishnoi.com
Website: www.monikabishnoi.com
LinkedIn: www.linkedin.com/in/monikabishnoi

CHAPTER 6

UNLEASH YOUR POTENTIAL: THE POWER OF AI IN EDUCATION AND TRAINING

By Fabian Bocek, PhD
Innovation and AI Expert
Frankfurt, Germany

> *Education is the most powerful weapon which you can use to change the world.*
> —Nelson Mandela

Today, almost everyone has access to interest-based courses and tools from anywhere, which has led to a growing demand for non-traditional education. Learners are increasingly turning to these individualized learning opportunities. They are seen as more attractive and practical than traditional qualifications. The industry is responding to this trend by focusing on specific skills and competencies rather than formal educational qualifications. These changes signal the need to develop more flexible, competency-based pathways that meet the demands of today's economy and tomorrow's world.

The use of artificial intelligence (AI) in education and corporate training is transforming the way learners acquire knowledge and develop skills. AI is now moving beyond the classroom to become a constant companion that supports and encourages students, employees, and anyone else on their learning journey. Institutions are developing more sophisticated AI capabilities that not only improve the learning process, but also act as personalized assistants that continuously support learners.

There is a wide range of applications for AI technologies. However, to unleash their full potential, this chapter illustrates the following four key approaches to AI in education and training: learning analytics, adaptive learning, assistant systems, and content creation. These technologies enable a personalized educational experience where AI supports learning in a variety of ways, from analyzing and predicting the learning behavior, to interacting with virtual tutors and chatbots, to creating specialized content.

This chapter promotes the individual potential for AI-based learning while demonstrating that it is flexible and accessible to everyone, regardless of academic background or geographic limitations. However, it is also important to understand that governmental and non-governmental institutions have an important role to play in unleashing the power of AI in education and training. Therefore, we will first look at the role of the state's ability to regulate AI and foster innovation.

Government Capacity to Regulate AI

Government institutions play a central role in advancing AI technologies. They can support the development of AI skills through a variety of policies, for example, promoting AI-based education platforms and lifelong learning initiatives. Innovation policy has the potential to strengthen the education and research sectors and encourage public-private partnerships. Research institutions and companies can support new innovations in AI through research funding, such as the Central Innovation Program for Small and Medium-Sized Enterprises (ZIM) in Germany.

Governments can regulate AI and foster innovation by enabling people to benefit from new AI developments. Collaboration and a good relationship between public resources and private individuals are crucial when it comes to acquiring skills, knowledge and expertise in the field of AI. The quality of AI policy depends on the relational nature of state capacity. State capacity allows for approaches that consider culture and human agency. The key to promoting AI, therefore, lies in the relationship between the state and its citizens.

The transformative capacity of a state to support new AI applications depends on its capacity to raise the education and skill levels of its population. This means that mastering the full potential of AI depends on individual human action. Here are four approaches to AI that can unleash your full potential of education and training:

Four Approaches of AI in Education and Training

AI is revolutionizing education and training by providing innovative solutions to improve learning outcomes. The transformative potential of AI is realized by four key approaches. Learning analytics uses data to understand learning behavior, assess performance, and provide automated feedback. Adaptive and personalized learning tailors educational experiences to individual needs. Assistant systems and chatbots provide real-time support, guiding learners through problem-solving and skill development. AI-driven content creation enables the customization of learning materials, making education more dynamic and responsive to learners. Together, these four approaches open up new possibilities for growth and learning.

1. Learning Analytics

Learning analytics (LA) is an increasingly important area in the education landscape that deals with the collection, analysis, and evaluation of data generated during the education and training process. The main goal of LA is to gain insights into learning processes and behavior in order to improve them in a targeted manner. AI is used as a key technology to better organize and analyze the large amounts of data that are generated from various sources. The data can then be applied for decision-making.

Learning analytics involves interpreting data from various learning activities to provide insight into individual learning behaviors. Whether for students, employees, or anyone seeking personal growth, learning analytics helps identify strengths, challenges, and progress. Using AI-driven tools, it can recommend strategies to improve the learning experience and track improvement over time. It empowers learners to take control of their education and helps them stay focused on achieving their goals.

In this way, learning analytics improves the quality of education and training programs.

Learning analytics plays an important role in vocational training. Organizations use these technologies to monitor the learning progress of their employees and optimize their training programs. By analyzing performance in various training modules, corporations can take targeted measures to improve the effectiveness of training and ensure that employees are acquiring and continuously developing the skills they need.

For educators and trainers, it offers valuable feedback on how learners are engaging with content, which allows for effective support. The analysis of data leads to individualized strategies to improve the learning experience. In essence, learning analytics is the key to personalized learning, providing the tools to continuously improve each individual's learning journey and help them achieve their goals more effectively.

Examples for Learning Analytics

- **Retention Analytics Dashboard (RAD):** RAD shows how combining different types of data can lead to a more nuanced understanding of a student's performance. **This enables support staff to be proactive and focus on the most at-risk students for early support. It aligns with the goal of learning analytics to monitor and actively improve student outcomes.**

- **FEATuring You:** developed by Southern New Hampshire University, FEATuring You is a digital assessment system that uses AI to evaluate and develop soft skills. The system aims to improve employment and educational opportunities for young people. Funded by Google.org, it assesses six key soft skills—communication, critical thinking, customer service, adaptability, results orientation, and problem- solving—through several assessments. The platform offers training and digital badges to help people learn new skills.

2. Adaptive and Personalized Learning

Adaptive and personalized learning refers to intelligent teaching methods that aim to adapt learning tasks and resources to best meet the individual needs of learners. These personalized approaches consider skills, competencies, and expectations to create a customized learning experience. Through AI-based learning tools, learners interact directly with technologies that serve as companions on their learning journey. They enable learners to master tasks more easily and to create a more efficient learning experience.

Adaptive learning refers to systems and technologies that dynamically adapt the learning process to the learner's current level of understanding and progress. An adaptive learning system continuously analyzes the learner's performance, identifies weaknesses or strengths, and adjusts the difficulty level and content of the learning material accordingly. It reacts in real time to the learner's actions and offers specific support or additional exercises to maximize learning progress.

Personalized learning goes beyond adapting to a learner's level of understanding and considers a broader range of individual needs and preferences. It involves adapting content and learning methods not only to a learner's ability, but also to their interests, learning goals, environment preferences, and pace. Personalized learning can involve different aspects of the learning process, including the choice of learning resources, the sequence of topics, or the type of activities that best suit the learner.

There are several approaches, tools, and platforms that support adaptive learning. Recommendation systems are used in digital learning environments to provide learners with the right offers and further learning activities. They are tailored to the specific needs and preferences of the user. Intelligent tutoring systems refer to programs that support individual learning by focusing on the user's performance. It aims to provide knowledge in a specific area by analyzing the user's skill level and providing tailored learning content based on that.

Learning management systems such as Moodle and Blackboard are software platforms designed to manage, deliver, and track education or training programs. They are used in schools, universities, and workplaces for online courses, training programs, and hybrid learning approaches. Massive open online courses (MOOCs) such as edX and Udemy are digital platforms that offer a wide range of courses that are typically open to anyone. Often free or with a low-cost certification option, they cover topics ranging from academic disciplines to professional skills.

Examples for Adaptive Learning and Personalized Learning

- **Knewton Alta:** Knewton Alta is an adaptive learning platform designed to personalize the educational experience. Designed primarily for higher education, Alta integrates with course materials to customize content based on individual learner needs. The system dynamically adjusts the difficulty of assignments and provides real-time feedback, ensuring that students master basic concepts before moving on to more advanced topics. By continuously analyzing student performance, it provides personalized pathways to help learners overcome specific challenges and achieve better results.

- **Coursera:** Coursera is an online learning platform that offers a wide range of courses in various fields, making quality education accessible to learners worldwide. It provides opportunities for students, professionals, and lifelong learners to learn new skills, earn certificates, and complete degrees. The platform collaborates with universities and organizations to deliver content that is state of the art. Coursera also applies learning analytics and personalizes the learning experience by recommending courses and materials based on individual performance and specific needs.

3. Assistant Systems and Chatbots

Assistant systems and chatbots are transformative tools in education and training, designed to enhance the learning experience by offering personalized, real-time support. These AI-driven systems vary from basic automated responses to advanced interactions that closely emulate human conversations. They are particularly valuable for their ability to provide personalized learning support, guiding learners through content, and recommending resources tailored to individual progress.

The tools are always accessible, allowing learners to seek help when they need it, which is especially beneficial in self-paced learning environments. In addition, these systems are highly scalable and can serve large numbers of users simultaneously.

Writing assistance tools are a subset of these systems that help learners to improve their writing skills. These tools provide real-time feedback on grammar, style, and content. Advanced systems can adapt to a user's writing style over time and offer personalized advice. AI-based writing assistants provide learners with immediate feedback and improvements. Tools like Grammarly or DeepL Write assist foreign language learners by providing corrections, suggestions, and sentence completions. In addition, users can compare their foreign-language texts with their native-language counterparts simultaneously, which enhances their understanding of the new language and helps them develop creative writing skills.

Modern chatbots are designed to respond to emotional cues. They can offer encouragement to learners who feel frustrated or demotivated. While these tools do not replace human interaction, they contribute to a more empathetic and supportive learning environment.

Assistant systems and chatbots are used in a variety of educational contexts, from primary and secondary education to higher education and corporate training. In schools, they assist younger students with homework and reinforce classroom learning. In higher education, they support research, give career advice, and help with administrative tasks such as registration and course selection. In the corporate world, they facilitate onboarding processes, provide continuous professional development, and assist with compliance training. Overall, they are revolutionizing education and training by making learning more engaging and effective.

Examples for Assistant Systems and Chatbots

- **ChatGPT:** OpenAI's ChatGPT is a sophisticated AI chatbot designed to engage in human-like conversations. It uses natural language processing to understand and generate text, which makes it valuable for tasks such as answering questions, providing explanations, and assisting with creative projects. As a virtual tutor, ChatGPT offers personalized support and helps learners to navigate complex topics. Its adaptability and broad range of applications position it as an effective tool in the evolving landscape of digital assistants and chatbots.

- **Automated Feedback:** the AI-powered application offers automated feedback for academic writing through a collaboration between FeedbackFruits and two universities in Rotterdam. Integrated as a plug-in within a learning management system, it delivers formative feedback including corrections, comments, and suggestions. This tool helps students to gradually improve their writing skills while allowing teachers to gain more time to provide in-depth feedback.

4. Content Creation

The field of content creation is emerging as a transformative force that directly impacts learners by making educational experiences more personalized and efficient. Unlike traditional content generation, AI-driven tools use advanced algorithms to quickly produce a wide array of educational materials. This gives learners access to extensive resources that are customized to their individual needs and styles.

A key feature of AI-driven content creation is its capacity to adapt to each learner's unique abilities. By analyzing data on an individual's progress and preferences, AI can generate content that addresses specific areas where the learner may need additional support. This approach aligns with the principles of adaptive and personalized learning, which aim to deliver the most relevant material to each learner.

Integrating content creation with learning analytics ensures that the generated materials are not only broad in scope but also relevant to the learner's current level of understanding. By continuously analyzing how learners interact with content, AI systems can refine and adjust the material, making learning more dynamic and responsive. This creates a feedback loop, in which the content continuously evolves to better meet the learner's need, thereby optimizing their educational journey for maximum effectiveness.

In addition to delivering personalized content, AI-driven content creation tools are capable of creating interactive learning experiences. These tools can generate multimedia content such as videos, animations, and simulations that make complex concepts more accessible and easier to grasp.

Examples for Content Creation:

- **Jasper:** Jasper is an AI tool designed to aid in content creation, particularly in writing. Utilizing advanced algorithms, it generates text based on user input, making it useful for crafting articles and blog posts. With a range of templates and customizable features, Jasper helps users to produce contextually relevant content efficiently. It increases productivity by reducing the time and effort needed while maintaining consistency in branding and messaging. Its flexibility makes it a useful resource for both content creation professionals and newcomers.

- **Canva:** Canva is an online platform that allows users to create a wide range of visual content, including presentations, social media graphics, and posters. Featuring an intuitive interface and an extensive library of templates, Canva caters to both professional designers and those with little design experience. It offers additional design elements and AI-driven tools that suggest layouts and styles to improve the creative process. Whether for personal projects or professional work, the platform allows users to create polished designs, making it a popular choice for those needing to generate visual content quickly and effectively.

Conclusion: Unleash Your Potential with the Power of AI

AI-powered tools like DeepL Write, Jasper, and Grammarly have demonstrated significant value in education and training, but they also pose challenges, such as encouraging passive text adoption without deeper cognitive engagement. The integration of AI into the learning process should complement, not replace, the human aspects of education. While chatbots can handle routine queries and provide personalized content, the critical thinking and deep understanding that results from human-led discussions and mentorship are irreplaceable. AI can provide basic support, but human educators remain essential for guiding learners through more complex cognitive tasks and fostering meaningful intellectual growth.

The four AI approaches illustrated are interconnected and mutually reinforcing, creating a comprehensive learning experience. Learning analytics generate insights that power adaptive learning systems and ensure personalized learning paths. Assistant systems offer real-time support based on data-driven recommendations. AI-powered content creation provides tailored and dynamic learning materials that suit individual needs. By effectively integrating these approaches, AI empowers learners to realize their full potential, overcome challenges, and achieve their educational and training goals with greater efficiency and depth.

About the Author

Dr. Fabian Bocek is a German expert in innovation and artificial intelligence. He holds a PhD in political science and a diploma in business administration. Since 2021, he has been working at the RKW Competence Center, a leading German think tank, where he is directly commissioned by the Federal Ministry for Economic Affairs in Berlin. As the head of the Innovation Promotion Program, he oversees research publications within the framework of the Central Innovation Program for Small and Medium Enterprises (ZIM).

His recent monograph, published by Routledge (Taylor & Francis Group), offers a comparative analysis of innovation and industrial policies in South and Northeast Asia, with a focus on the automotive and information and communication technology sectors. Dr. Bocek's professional and academic interests span innovation, digitalization, and artificial intelligence; international and comparative political economy; innovation, technology, and education policy; public sector analysis; and economic transition and development.

Email: fabianbocek@gmx.de
LinkedIn: https://de.linkedin.com/in/dr-fabian-bocek

CHAPTER 7

DIVERSITY, ENTREPRENEURSHIP, AND AI: DRIVING INNOVATION IN THE GIG ECONOMY

By Fabio Brand
Founder, Sagui.AI, AI-powered Go-to-Market Platform
Rio Grande do Sul, Brazil

> *The future of AI is to empower every person and every organization to achieve more.*
> —Satya Nadella, CEO of Microsoft

While artificial intelligence (AI) has been a common topic in the tech industry for over a decade, its applications in the business world began to take shape exponentially around 2021. Just a few years ago, the idea of machines performing tasks, interacting, and learning seemed like something out of a futuristic Hollywood movie. Today, practical applications of AI are rapidly developing. AI now plays a significant role in businesses, but unlike what was portrayed in movies like *Blade Runner* or cartoons such as *The Jacksons*, it isn't here to replace people—it's here to empower them. It's inevitable that some highly operational tasks will be automated, allowing companies to create more efficient environments that positively impact business results. However, by automating repetitive tasks, AI helps businesses save time and resources, enabling employees to focus on more strategic and creative activities.

Entrepreneurship and AI are converging to create new opportunities for innovation at a truly rapid pace. This transformation is redefining how we work, create, and thrive in an increasingly digital and interconnected world.

The Power of Diversity in Entrepreneurship

Diversity isn't just a social concept; it's an essential driver of innovation. Businesses that embrace diversity in their teams and practices are often more innovative and better positioned to respond to market changes. Diversity of thought and experience enriches the creative process, allowing companies to visualize challenges from different perspectives.

Early in my career, I realized that diversity was a competitive advantage. As an LGBTQIA+ person in one of the most economically unequal countries in the world, working with people from different backgrounds, stories, and skills not only brought me new perspectives but also strengthened my resilience and adaptability—traits that would become fundamental to the success of my business in the future. Coming from a humble background and being the child of divorced parents, I always knew what it meant to manage a budget with limited resources, a lesson I learned month after month watching my mother balance the family budget with two jobs. Often, to supplement her income as a nurse, she would do what we call "*bicos*" in Brazil—gigs—working a few hours at a pizzeria or selling cosmetics to friends and colleagues, creating a mutual support network where people in economically tough situations would support each other and consume each other's products or services, sometimes even paying more. In the future of work, where freelancers and individual entrepreneurs play increasingly active roles, diversity will become even more essential in creating solutions that meet a wide range of increasingly specific needs and opportunities.

As I took my first steps as an entrepreneur, the main challenges I faced were the lack of resources to build a consistent online presence and the difficulty of hiring a team before having a stable revenue stream.

Trying to wear multiple hats at once, I noticed that some tasks ended up being left behind and took time to be picked up again, resulting in lost opportunities. Now, with the power of AI, a solo founder can create their first website, produce content, schedule posts, manage leads, and automate repetitive tasks, becoming more productive and capable of generating revenue until it's time to hire a team. This process of achieving entrepreneurial success can indeed be shortened with the use of AI.

AI is one of the most transformative forces of our time. With its automation capabilities and predictive analysis, AI offers small businesses the ability to compete at a level that was previously out of reach. It not only improves operational efficiency but also allows entrepreneurs to focus on more strategic activities. The ability to process large volumes of data and uncover hidden patterns provides a competitive edge by enabling more informed and data-driven business decisions.

An example is how AI is being used to automate repetitive processes. In customer service, for instance, through conversational AI and chatbots, it's already possible to create true "brains" that learn and respond to questions in sales or support. Additionally, AI can learn from human reactions, making service increasingly efficient. Efficiency and AI are key words when it comes to managing limited resources in the business world. This frees up time for entrepreneurs to focus on more strategic tasks, such as product development and customer experience personalization.

Predictive analysis with AI also allows businesses to anticipate trends and make informed decisions, adapting their strategies to better meet consumer demands. This is especially valuable for small business owners, where agility and adaptability are crucial for success. This is a true example of access to technology that was once very expensive and only available to large corporations.

Another application lies in marketing optimization and automation, enabling A/B testing as well as understanding customer behavior. The application of AI in small businesses can reduce costs and improve operational efficiency so that small companies can be as competitive as larger ones, with a leaner structure.

One last example is the ability to establish a professional online presence in seconds. A task that once required technical skills like programming, marketing, or design, taking weeks or even months and involving many steps, can now be done by creating an entire website with text and images in seconds, or by generating content and designs for social media from a simple prompt. Of course, solutions like these will currently need curation and a human touch, but it's very exciting to think about what small businesses will be able to do in the next 10 years.

Entrepreneurship in the Gig Economy

The gig economy is rapidly growing, driven by technology that allows individuals to work independently and flexibly. For many, this represents an opportunity to escape the constraints of traditional work, avoiding repetitive tasks and following their passions, having more time with family and friends, and finding more challenging work. For professionals seeking more flexibility, autonomy, and financial return, the gig economy represents an alternative to conventional work and the 9-to-5 cubicle.

However, the gig economy also presents unique challenges, such as the lack of financial stability and traditional benefits. In this activity, AI can play a crucial role by providing tools that help freelancers manage their finances, optimize their time, and expand their business opportunities. Financial and time management can be significant challenges for those who work anytime, anywhere.

The Future of Work: Entrepreneurship, Flexible Work, and Personal Investment

One of the main areas where technology can have a significant impact is the job market. As the future of work advances rapidly, we see companies streamlining their operations and talents seeking new opportunities with more flexibility. Entrepreneurship emerges as a window of opportunities, especially in the United States, where a record 5.5 million new businesses emerged in one year. In Brazil, the MEIs (individual microentrepreneurs) are reshaping the economic landscape, representing 60% of all formalized businesses.

This doesn't mean that everyone wants to be an entrepreneur, but the reality is that companies are becoming increasingly lean, seeking efficiency with fewer resources. This requires people to learn how to monetize their skills and work with multiple clients, creating diversified income streams.

Investing in yourself has never been more crucial. In a constantly changing job market, developing new skills and being prepared to reinvent yourself quickly is essential. Digitization and the use of technologies are further expanding opportunities, allowing professionals to reach new markets and expand their businesses from the comfort of their homes or from a chair on a beach in the Bahamas.

A New Path: Portfolio Careers

It is estimated that 50% of professionals will have a portfolio career by 2030. The creative economy is expected to double in size, reaching $480 billion in the next five years. This represents an opportunity for professionals to diversify their income streams and explore new ways of working.

Tips for staying competitive include investing in yourself, diversifying your income streams, learning to monetize your knowledge, and promoting yourself in the market. Considering having a *"bico"* to complement your income while maintaining a formal job can help build financial security, generate a portfolio, and expand your connections. Remember when I mentioned that my mother was doing this 20 years ago? It was the "gigs" that helped her buy her first financed apartment. Even though she had a formal job, she always explored gigs as a significant complementary income source.

The AI Revolution in Small Businesses

AI is democratizing access to advanced tools, enabling even the smallest companies to reach levels of innovation that were previously unattainable without significant funding. This is particularly important for entrepreneurs in less privileged regions, where access to funding is limited.

Small businesses that adopt AI can compete with large companies, using automation to optimize their operations and data analysis to better understand their customers. This not only improves efficiency but also allows them to offer more personalized products and services. AI, as a powerful tool, can increase efficiency and profitability, but it requires careful and ethical implementation to ensure that companies position themselves ahead in the market.

Moreover, AI is helping overcome language and cultural barriers, allowing people and businesses to expand their reach. With simultaneous translation and cultural analysis tools, companies and professionals can adapt their messages and products to meet the needs of consumers in different regions. For example, as someone born in Brazil, where only 1% of the population speaks English fluently, I can write this chapter in Portuguese and rely on AI to make the writing flow more naturally for a North American reader. AI is not replacing me or writing for me; it's enhancing a skill I still need to develop.

This reduces acquisition costs, making small businesses more competitive and profitable. For small businesses and entrepreneurs, AI offers an opportunity for continuous innovation. By leveraging technology to automate tasks and gain data insights, they can focus on creating value and differentiation in the market, doing more with much less.

As AI continues to evolve, it's important for entrepreneurs and small businesses to take a proactive approach to its implementation. This means investing in training and skill development to ensure that teams are prepared to use AI tools effectively and productively.

Entrepreneurs should see AI not as a threat but as an opportunity to amplify their capabilities and create new forms of value. By integrating AI into their operations, they can position themselves as leaders in innovation, ready to face challenges and seize opportunities in the future.

Final Message

We are in an era of unprecedented opportunities, where diversity, entrepreneurship, and artificial intelligence are converging to create a more democratic, inclusive, and equitable future for success. By embracing this technological revolution, we can not only improve our operations and offer better products and services but also contribute to a fairer world.

The future of work is collaborative, where humans and technology work together to create solutions that meet the needs of a constantly changing world. This chapter is an invitation for all entrepreneurs and small businesses to embrace AI as a partner in their entrepreneurial journey. This is the AI Mindset.

About the Author

Fabio Brand is a Brazilian entrepreneur who has dedicated his professional career to creating value and helping small businesses thrive digitally through his marketing agency. He holds degrees in business administration and marketing, a postgraduate degree in digital influence, content, and strategy, and is currently studying software engineering and AI. In 2024, he founded Sagui.AI, an AI-powered go-to-market platform, aimed at creating opportunities for gig economy professionals and small business owners to thrive online.

Email: fabio@sagui.ai
LinkedIn: https://www.linkedin.com/in/fabio-brand/
Website: www.sagui.ai

CHAPTER 8

TRANSFORMING CUSTOMER EXPERIENCE WITH AI— STRATEGIES AND INNOVATIONS

By B. A. Marbue Brown
Founder, CEO, Author
New York Metropolitan Area

> *AI will be the most transformative technology of the 21st century. It will affect every industry and aspect of our lives.*
> —Jensen Huang, CEO of NVIDIA

Perhaps one of the greatest opportunities for AI to make a difference in society is in transforming the experience that businesses and institutions deliver to their customers across the entire value chain from discovery to shopping to purchasing and servicing. This transformation doesn't necessarily portend an entirely new set of approaches or strategies to radically improve customer experience (CX), at least not immediately. Instead, it is recognition of the fact that AI has enabled businesses to crack the code to deliver exceptional experiences through approaches that have long eluded their grasp due to the limitations of prior technologies.

The arrival of generative AI (GenAI) with its amazing conversational, problem-solving, content creation, and artistic capabilities has been a game changer that unlocks opportunities for businesses to engage with customers in ways that they wanted to but hadn't been able to previously. Of course, it also opens up possibilities to elevate experiences for customers in ways that were previously unimagined.

In this chapter, we will delve into specific examples of how AI is removing barriers to enable breakthrough CX enhancements and look ahead to next generation CX enhancements that may be on the horizon as AI continues to gain capabilities. There are at least eight perennially intractable problems that AI is helping businesses and institutions to solve. Let's start by digging into these first.

Extract Key Takeaways from Unstructured Customer Interactions

One of the greatest challenges for businesses is to extract insights from the myriad of the unstructured interactions they have with their customers daily. That includes insights from customers' conversations with customer service associates (CSAs) by phone, chat, or email, as well as from posts and reviews left by customers on social media. These interactions contain a wealth of information about overall customer sentiment, brewing threats, as well as untapped opportunities for invention/innovation. The problem that has eluded companies for years on end is how to parse this information, which is stored as freeform text, audio files, or video clips.

Numerous attempts have been made to address this challenge using a variety of text analytics models, and results have been passable but have also still left a lot to be desired. Meanwhile, GenAI's advancements in natural language understanding (NLU) have ushered in breakthroughs in our ability to parse this type of data to extract themes and sentiment as well as to synthesize key takeaways from the content to pinpoint red flags issues that need to be addressed urgently or innovations that could be pursued to address unmet customer needs. ChatGPT's exploits have provided ample evidence of how AI has advanced the state of the art for NLU. Now businesses can put those NLU advancements to work for them.

Armed with the information captured using AI-powered models, businesses don't have to wait to ask customers about their contacts to learn what's working, what's not, or what's on customers' wish lists. They can extract that from conversations in near real time and synthesize it into insights and actions that deliver benefits for customers almost immediately.

Simplify Navigation of Complex Mobile Applications

Another major challenge that businesses have grappled with is migrating customers to mobile app usage, particularly customers from generations that aren't "born digital." Many companies have taken an "if you build it, they will come" approach, only to find that the customers aren't coming, and the reason they aren't is quite simple, i.e., they aren't confident that the app has the capabilities they need to accomplish their intended tasks, and if those capabilities are in the app, customers don't know how to find them. AI is radically changing the landscape of app navigation by providing a conversational interface for digital assistants that guides customers through discovering mobile app features as well as navigating mobile app workflows to accomplish tasks.

For example, when using a banking app, instead of browsing around to find out how to perform certain functions like how to send a wire transfer or how to retrieve annual tax documents, customers can simply speak to the app or type in a natural language question on the home screen to navigate directly to the appropriate screens. The AI-enhanced digital assistant can then guide them through the process of accomplishing the specific task they had in mind. Similar capabilities were available prior to GenAI, but AI has significantly expanded the number of functions that can be accessed this way, the effectiveness of the queries to access them, and the success rates for customers achieving their ultimate objectives.

The enhancements delivered by GenAI will make customers feel like they have their own personal assistant to help them learn how to use the app and navigate their way through it. In turn, that will remove key barriers to digital adoption and accelerate the pace.

Reimagine IVRs to Fast-Track Resolution of Inquiries

Navigating interactive voice response (IVR) systems has been a perennial problem as well as an irritant for customers. No one wants to go through a maze of questions to get to the correct agent to address their inquiry or, worst yet, wind up at a dead end without getting the answers or resolution they need. Yet, that's been the typical experience for customers calling into a business.

GenAI is changing all that by adding significant new capabilities to IVR systems that enable them to better resolve customers' inquiries directly or to more easily route customers to agents that can help them. Now customers can get straight to the right agents by simply describing what they would like to accomplish versus going through a series of prompts.

For example, an AI-enabled IVR system could easily capture all of the information necessary to file an automobile accident insurance claim, then email the populated claim form to the customer for final review and submission without engaging a live agent. Also, an AI-enabled IVR could capture the information needed to schedule follow-up visits or routine check-ups at doctors' offices. Banks are using AI-enhanced IVRs to respond to routine inquiries to find out about account balances, whether certain checks have cleared, or when funds will be available after a deposit.

Transform Chatbots into an Attractive Option for Resolving Inquiries

GenAI is being used to enhance chatbots to seamlessly handle more customer inquiries directly without plunging them into infinite loops that frustrate them instead of solving issues or to quickly transfer them to live agents who could. AI-enhanced chatbots are being designed to detect words and phrases that imply urgency, frustration, confusion, or other sentiments that would trigger a transfer to a live agent. They are also being designed to detect complexity that would automatically warrant assistance from a live agent. Meanwhile, for inquiries that a GenAI enhanced chatbot can handle, the conversations are being upgraded to reasonably simulate a chat conversation with a live agent to the point that customers can hardly tell the difference. Chatbots are even being designed to deliver surprise and delight moments to customers based on the nature of their inquiry.

Minimize Time to Onboard New Customer Service Associates (CSAs) with Proficiency That Rivals Tenured Agents

Two of the biggest challenges for contact centers include 1) the lead time it takes for new agents to reach the optimum proficiency to begin taking live contacts with customers and 2) the disparity between the experience delivered by veteran agents versus newbies. GenAI has been effectively deployed to help solve both of those by simulating live call scenarios that agents can use to hone their skills during training.

These AI-powered simulations enable new agents to get acclimated to real-world conditions in an environment where it's safe to make mistakes and that enables them to rapidly run through as many repetitions of specific scenarios as are necessary to gain high levels of proficiency before they transition to taking live calls. This simultaneously helps to narrow the gap between new agents and experienced agents, enabling customers to get a more consistent experience regardless of the tenure of the agent handling the contact. Meanwhile, even tenured agents can use the AI-powered simulations to continuously upgrade their skills to gain mastery of the most difficult contact types, so they can deliver more seamless experiences for customers.

Previously solving these challenges was seemingly intractable and much more expensive because it required pulling tenured agents or supervisors offline to role play with new agents to help get them ready. It simply was not feasible to pull enough personnel offline to do role plays with a full class of new agents considering that role playing is a one-on-one activity. Additionally, even for a limited number of new agents, it was not feasible to pull personnel offline long enough to work through the number of iterations required to develop peak proficiency or to work through the variety of scenarios new agents needed to be ready to handle.

AI-powered simulations have put these issues in the rearview mirror, and some well-known brands are already reaping the benefits of deploying this technology, including Sallie Mae, Western Union, and some major banks, all of whom deployed a solution delivered by Zenarate. On the strength of that solution, Western Union has reduced agent time to proficiency by more than 50% and improved CSAT scores by 33%. Sallie Mae has reduced agent certification time for live calls by more than 75%.

Eliminate (or Minimize) Lag Time for Agents to Research Answers to Customers' Inquiries

Another key customer experience challenge that contact centers have struggled mightily to solve is the lead time it takes for agents to research a customer inquiry before they can provide an answer. Agents frequently have to put customers on hold while they research answers to questions, and they often have to dip into multiple systems to get the information needed to give an appropriate response. That process can take several minutes while customers wait in limbo to get a resolution to their inquiry, which is not a positive experience for them.

GenAI has been deployed with great effect to help solve this problem by training LLMs on companies' internal systems and knowledge bases to retrieve answers for customers' inquiries in seconds versus minutes. Datamark, a company that supplies solutions for this use case, estimates that the time savings range from 30 seconds to two minutes and that this has a favorable impact on both the employee and customer experience. End users who have deployed GenAI for this use case include El Paso 311 and PDI Technologies (administrator for Shell Fuel Rewards).

Offload Low-Complexity Inquiries to Chatbots or Virtual Assistants

Perhaps one of the areas that has stirred up the most controversy is whether GenAI solutions are poised to replace live contact center agents outright. Concerns range from wholesale job loss in contact centers to potential negative impacts to CX from GenAI applications that can't deliver experiences on par with live agents. Meanwhile, some companies have taken the plunge and deployed GenAI solutions to handle customer inquiries directly and have achieved favorable outcomes.

For example, Toyota's Destination Assist service enables drivers who have factory-installed navigation systems to download directions to their vehicles at the touch of a button. Previously, when Destination Assist was activated, drivers would speak with a live call center agent who would take their request, then transmit directions back to their cars. Following the rollout of an AI-powered version of Destination Assist in 2023, the automated agent now handles 95% of requests, which frees up live agents to handle more complex requests.

Another example is Klarna's in-house developed "AI Assistant" that handled about two-thirds of its customer services conversations over the course of a year from 2023 to 2024 while covering a wide range of contact types (e.g., refunds, returns, resolution of invoice inaccuracies) and achieving customer satisfaction scores on par with live agents. AI Assistant has helped Klarna reduce repeat inquiries by 25% and cut the time to serve by more than 80%, both of which are indicative of better experiences for customers.

These are significant developments because prior deployments of technology to offload routine inquiries produced uneven results and regularly left customers frustrated. Now businesses have a viable path to offload repetitive calls that delivers great CX and supports agents by only sending them customers that really need their help and intellect.

Summarize Customers' Prior Contact History

One of the most frustrating experiences for customers is being forced to retell their story from scratch to one CSA after another. Despite all kinds of attempts to mitigate this issue, businesses have struggled mightily to equip their agents with background information that familiarizes them with a customer's most recent touchpoints prior to the current contact as well as other pertinent information about the customer that could be used to personalize the contact. That info would enable agents to give customers the feeling that "this company gets me" and that "this company has a clue about what I've been through to get to this point."

GenAI is ushering in a new breed of copilots that can poll backend systems to bring forward salient information from the customer's most recent touchpoints, especially ones that did not result in the customer completing their objective. With that info, CSAs will be able to help customers pick up where they left off instead of asking them to tell their story from scratch or to answer questions that the CSA can answer with information that the company previously captured.

These copilots can also bring forward information about the customer's relationship with the business such as purchase history, contact history, returns history, feedback history, etc. In a nutshell, these copilots deliver a 360-degree view of the customer that businesses have struggled to achieve for so long. As such, GenAI copilots that are trained on the data sources CSAs use to bring themselves up to speed during customer contacts will be invaluable for enabling more personable contacts and raising the bar for CX. Sales, business development, and other roles will also benefit from these copilots.

Transforming CX with Next Generation Use Cases Powered by AI

The breakthroughs provided by AI to crack the code on these eight previously intractable challenges represent a quantum leap for CX. Meanwhile, the opportunity to transform CX with AI is not limited to scenarios that were already on businesses' radar. A whole new world of previously unimagined use cases has opened up, and businesses are stepping up to take advantage of them.

For example, in early 2024, Amazon started rolling out "Rufus," its GenAI-powered chatbot that helps shoppers make decisions about which items to buy, especially when they don't know which features are most important to consider when making the purchase decision. For example, you can consult Rufus to help you figure out which big screen TV to buy, what factors to consider when choosing a pair of earphones, or what items to buy to prepare for an upcoming trip. In response, Rufus can offer recommendations depending on whether you're focused on price, popularity, or particular use cases (e.g., best earphones to use for Zoom meetings, listening while working out, or playing with a band). Rufus can even suggest and answer follow-up questions to help you narrow down your choice. The chatbot can do this because it is trained on the Amazon product catalog, Amazon customer reviews, and Amazon community Q&A responses, as well as other data from across the web.

Meanwhile, Sam's Club has rolled out its AI-powered answer to stationing "receipt checker" persons at the exits to its warehouse club stores. That solution worked so well to reduce the exit bottleneck in the first 20% of its stores where they deployed it that by the time this book goes to publication, it will already be deployed widely across their network.

Besides those already mentioned, some companies have deployed bespoke GenAI solutions to score cell phone values for trade-in, to translate community responses in technology user groups into multiple languages, to provide closed captioning for accents that live agents have trouble understanding, and more.

The bottom line is that there's no shortage of opportunity for businesses to benefit from deploying AI to transform CX, and for many that transformation is well underway. If your executive team is not sure where or how to get started with AI, try shifting your vantage point from "Where should we start with AI?" to "Which CX use cases that we are already trying to solve would be tackled more easily if we apply GenAI?" In a nutshell, start with the use case not the technology, and the path forward will be illuminated very quickly. Once you've crossed that bridge and demonstrated business value, other opportunities will present themselves more readily.

About the Author

B. A. Marbue Brown is author of the book *Blueprint for Customer Obsession* and the Founder and CEO of The Customer Obsession Advantage (customerobsession.net), a firm dedicated to helping companies achieve extraordinary business results through customer obsession. He is an accomplished customer experience (CX) executive with a track record of thought leadership and signature business results at some of the most iconic companies on the planet, including JP Morgan Chase, Amazon.com, Microsoft Corporation, and Cisco Systems. His expertise has been featured in premier media outlets, including The Wall Street Journal, The Washington Post, USA Today, Fox Business, CNN Business, CBS News, Yahoo! Finance, Forbes, The Baltimore Sun, The New York Post, Wharton Business Daily, and KTLA-TV.

Email: founder@customerobsession.net
Website: www.customerobsession.net
LinkedIn: https://www.linkedin.com/in/marbuebrown/

CHAPTER 9

THE DISPLACEMENT DILEMMA

By Nigel Cannings
Founder, Intelligent Voice; NLP/Speech AI Expert
London, England, United Kingdom

> *The first thing we do, let›s kill all the lawyers.*
> —William Shakespeare (Henry VI)

In my career, I have been to some deep, dark places.

On only a few occasions have the metaphorical and physical collided, but any young City lawyer of a certain vintage will be transported back to their first M&A transaction, locked in an over-illuminated, under-ventilated sub-basement of an office block lined floor to ceiling with paper and files, armed with one simple instruction: "Read, then summarise."

Expertise is earned. There are no shortcuts. We hear that it takes 10,000 hours, or 10 years, of deliberate practice to become an expert. Not just showing up for 10,000 hours, actually grinding away to become better. Sometimes I feel like I did all 10 years in a solid week of no sleep, reading one standard contract after another, but that was the caffeine talking. But I did work hard for 10 years to become the best lawyer I could. And then I quit. Technology seemed like more fun ...

Two years ago, I was still probably a lot smarter than the average computer when it came to reviewing a contract, even though it is 20 years since I last practiced law. I could pick up the major points, see where the draftsperson was being a bit unreasonable, where they had snuck in a one-sided assignment clause, and I could make a fair fist at doing some redrafting.

Today, I think we are about even. In fact, I was so confident that we could use AI to help us detect unfair contract clauses, I started a website that does exactly that, perhaps ironically focusing on changes in privacy policies that allow companies to take user data and use it to train AI models.

In two years' time, 95% of all legal work will be capable of being performed by AI. This includes contract review, drafting of pleadings, legal research, and due diligence. It will even involve elements of strategy in conducting proceedings.

In five years' time, you will be able to have AI conduct hearings on its own, presenting well-engineered, beautifully researched arguments in a well-modulated voice of your choice, indistinguishable from a human, apart from its ability to work 24 hours a day.

I am not joking. The legal profession in its current form is dead. It just doesn't know it yet.

The Death of Expertise

I am going to stop picking on lawyers for a minute. The legal profession is an easy target not because people don't like lawyers, but because what modern AI is trained to do is exactly the same thing lawyers are trained to do. It's about analysing vast amounts of information, identifying patterns, applying rules, and making decisions based on established frameworks. However, this issue extends far beyond the legal field. Other professions, such as medicine, finance, and even creative industries, face similar challenges. AI's capabilities in diagnosing diseases, predicting market trends, or generating content all mirror the tasks traditionally performed by human experts. This convergence of AI's strengths with the core functions of various professions underscores the broader implications of AI's rise, putting pressure on many fields that rely on specialised knowledge and expertise.

How do we train our next generation of doctors if all current diseases can be diagnosed by machine? How does tomorrow's tax expert gain their experience, if the answers to every question are a simple Google-style search away?

It is not just in "skilled" industries that this is going to be a problem. We will see this happen faster in those areas that are highly labour-intensive, but which consist, in the main, of repetitive and easily replicable tasks, such as customer service, particularly call centres.

Okay, this stuff is not infallible: a machine learning model designed to detect Covid-19 was actually only detecting whether people were lying down or not. Lawyers in New York were censured for presenting motions that were wholly made up by ChatGPT.

And oh, how we laughed when the DPD chatbot was hacked to swear at a customer and to disparage the company it "worked" for. A handful of Luddites downed tools and breathed easy, happy that their jobs were safe from AI.

However, the move to replace first- and second-line support agents both online and in the call centre is happening now. Clearly, we need better guardrails in place to prevent some of the more egregious errors and hallucinations in these early systems, but they will be fixed. Speech technology has evolved to allow AI systems to natively understand and respond to humans without even the need to convert the speech to text before it is processed. Linked to a customer knowledge base and with a set of parameters to give the "agent" flexibility, the customer will not even know they are talking to an AI agent.

But there are going to be times when the AI system just can't help. If it is an unusual set of events outside of what it is trained for. What then?

What Then?

This is the crux of our issue. AI is fantastically good at looking backwards and generalising forwards based on what it knows. Given all of the training data in the world, it can make an incredibly accurate guess about what comes next, if the next thing has happened before, or at least something similar. What AI does very badly is adapt to completely new scenarios.

When AI encounters a situation it hasn't been trained for, it falters. Worse, it often guesses. This is where human expertise, with its ability to draw on intuition, experience, and creative problem-solving, becomes irreplaceable. A machine can predict outcomes based on past data, but it cannot innovate, think outside the box, or respond with empathy and ethical consideration when faced with a unique or unprecedented challenge.

In a few years, all of our expert call centre operatives will have retired or taken other jobs. All of the people who would have replaced them, working at the coal face dealing with difficult customers day in, day out, will have been replaced by machines. When the tough calls come in, who is there to answer them?

Imagine a doctor faced with a rare, never-before-seen disease. While AI can provide insights based on known data, it's the human doctor who must navigate the uncharted waters, combining clinical expertise, creativity, and ethical judgment to treat the patient. Similarly, in law, when an unforeseen legal issue arises that falls outside of established precedent, it's the seasoned lawyer who must craft new arguments, set new precedents, and push the boundaries of the law.

As AI continues to encroach on the territory of human experts, the role of humans will shift from performing routine tasks to handling the extraordinary, the novel, the ambiguous. This shift raises a crucial question: how do we continue to cultivate expertise in a world where AI handles much of the routine work? How do we ensure that future professionals gain the experience necessary to deal with the unpredictable and the unknown?

If we allow AI to dominate too much of our professional landscapes, we risk losing the depth of human expertise that is developed through years of grappling with complexity, making mistakes, and learning from them. The challenge we face is not just about integrating AI into our workflows, but about redefining the value of human knowledge and ensuring that it continues to evolve alongside machines.

The death of expertise isn't just a loss of jobs; it's a potential loss of our ability to think critically, to innovate, and to apply wisdom in a world that increasingly relies on algorithms and data. The question isn't whether AI can do what experts do—it's whether we can afford to let it replace the nuanced, adaptive, and creative thinking that only humans can provide.

Enter Jevons, Stage Left

The arguments about the impact of new technologies are not new. Every major technological innovation has winners and losers. In the 19th century, rapid industrialisation led to the worry that we would kill cottage industries, lose expertise, and throw workers onto the streets as the machines did all of the required work. However, the economist William Jevons noted that more efficient steam engines actually increased coal usage rather than decreased it, as the efficiency led to more goods being available at a cheaper cost, with greater demand in the economy, thereby increasing the use of coal. This is known as the Jevons paradox and has been used to try to explain the effect of every major technological advance since.

So, should we not panic? Let's go back to the legal example.

Lawyers have survived a lot of change. Mechanisation has never held them back. From quill, to typewriter, to word-processing to electronic discovery, lawyers have ended up making more money from mechanisation, not less. Since 1990, there are 72% more lawyers in the US alone. More productivity equals more billable hours.

But AI is not a productivity tool. It can be, but it is so much more. Word processors are dumb. Search engines are dumb. AI is smart, very smart, at least in those areas that require expertise. I could probably have had this whole chapter written in three minutes using ChatGPT. It would probably have read better as well. With less humour and personality, probably, but would my readers really have cared? Probably not if I were billing them by the hour to write it. And this is what clever lawyers (and other professionals) should be terrified of. If 95% of what you have trained 10 years to do can be done for 50 cents by a machine in five minutes, what is the point of you?

Adapt or Die

I started this chapter talking about deep, dark sub-basements. The answers to the death of expertise may lie in a different set of basements, those inhabited by the mainframes of the world's banking systems.

About 60 years ago, IBM adopted a brand-new programming language for its mainframes: COBOL. Over the next 20 years, much of the world's banking infrastructure was built on the back of these mainframes and, therefore, this programming language. And mainframes run the world. Expecting a benefits payment in the UK? That'll be a mainframe. In fact, there are reckoned to be 800 billion lines of COBOL in use in the world today.

At various points, as COBOL programmers got older and closer to retirement, there were stories that the world banking system would collapse due to the lack of people to maintain it. The answer? Pay more, and train more people. Two very simple steps: identify the coming problem, and incentivise people to meet the challenge. Can the same be done to replace the lost expertise in other industries?

In the case of AI's impact on industry, first, you have to work out what it is that you want your experts to be able to do. Yes, they need to be able to analyse and critique the work that AI produces. But that is a role that will diminish as AI will learn from that process. What they really need to be able to do is innovate, invent, and expand. How do we train people to do that?

No Train, No Gain

The solution to the displacement dilemma lies not in resisting the tide of AI, but in fundamentally rethinking how we cultivate expertise in a world where machines perform so many of the tasks that once defined human mastery. The future experts must be those who understand not just how to use AI, but how to push beyond its limitations—those who can innovate in areas where AI falls short, who can apply creativity, ethics, and deep critical thinking to solve problems that machines can only partially address.

Redefining Education and Training

To achieve this, our approach to education and professional training must evolve. Traditional methods of teaching—rote learning, memorization, and repetitive practice—must give way to curricula that emphasize problem-solving, interdisciplinary thinking, and adaptability. Future professionals should be trained to work alongside AI, using it as a tool to enhance their capabilities rather than as a crutch that absolves them of the need to develop their own expertise.

I am never going to be a doctor, no matter how much AI you throw at me, but there is a huge opportunity to train people more effectively, and more quickly, to become empathetic and efficient diagnosticians—so long as blood doesn't scare them—because fundamentally, in the medical field we will always need a person to deal physically with a patient.

We are also going to have to create environments that encourage on-going learning and adaptability. As AI continues to advance, the knowledge and skills that are relevant today may become obsolete tomorrow. Continuous professional development should become the norm, with a focus on fostering the skills that AI cannot replicate: creativity, empathy, strategic thinking, and ethical judgment. So we should hopefully see an increase in innovation, rather than seeing it stifled.

The Role of Human Judgment

While AI can process vast amounts of data and generate insights at a speed and scale that humans cannot match, it lacks the ability to understand context, to make value-based decisions, and to apply the nuanced understanding that comes from lived experience. This is where human judgment remains irreplaceable. Training the next generation of professionals to harness this judgment, while also understanding the capabilities and limitations of AI, will be critical to ensuring that we do not lose the essence of what makes expertise valuable.

A Call to Action

The displacement dilemma is not just a challenge for the professions, but for society as a whole. We must invest in education and training that prepares individuals to thrive in an AI-augmented world, ensuring that human expertise continues to evolve and remains central to our progress. We must also engage in thoughtful, ethical discussions about the role of AI in our lives, and about how we can ensure that technology serves humanity, rather than the other way around.

As we navigate this new landscape, the question is not whether AI will replace experts, but how we can adapt to ensure that human expertise remains vital and relevant. By embracing change and focusing on the unique qualities that make us human, we can ensure that expertise does not die, but instead, is reborn—stronger, more innovative, and more indispensable than ever before, with innovation at the fore.

So, we will not see the true death of expertise, rather we will redefine what expertise means. Will this lead to a fundamental reshaping of a lot of traditional roles? Yes, it will. But will it enable humans to have more fulfilling careers, focusing on what we enjoy and are good at? If it saves one other young person sitting surrounded by a mountain of paper, I certainly hope so.

About the Author

Nigel Cannings is an expert in speech technology and AI and its application in the legal and financial sectors, having spent over two decades pioneering technology-driven solutions in these fields. He has over a dozen patents to his name in the field of natural language processing, cryptography, AI, and speech processing. A former City lawyer, Nigel transitioned from law to technology, where he has been at the forefront of developing AI tools that help understand communications better and pioneered the introduction of GPU technology into speech recognition in 2014. He is the founder of Intelligent Voice, a provider of speech and NLP solutions to the financial sector. He also founded Terms Were Not Disclosed, a groundbreaking AI platform that detects unfair contract clauses and tracks changes in privacy policies, focusing on protecting user data. Based in London, Nigel is a thought leader on the impact of AI on professional expertise, and he continues to explore the intersection of technology and human intelligence. He is also the author of the book *The Displacement Dilemma: Navigating the Survival of Human Expertise in an AI-Driven World*.

Email: nigelcannings@gmail.com
LinkedIn: https://uk.linkedin.com/in/nigelcannings

CHAPTER 10

GENERATIVE AI AND SOFTWARE DEVELOPMENT

By Rodrigo Cantú Polo, MSc
Senior Software Engineer
Curitiba, Paraná, Brazil

> *AI is the new electricity.*
> — Andrew Ng

Software development is my passion. My professional life as a software developer has been to understand what needs to be done regarding some areas to automate and deploy a working system related to this automation. The whole process to accomplish it is usually complex and time-consuming. In 2022, OpenAI's Chat GPT came along among other GenAI technologies, making artificial intelligence available for everyone in a way that was not possible before. Regarding text generation, they generate code impressively well, and their impact on software development is huge. In this chapter, we will dive into a brief explanation of how GenAI and software development work and how GenAI can be used to help in it.

What Is GenAI?

Generative artificial intelligence (GenAI) is a specific subset of the artificial intelligence domain. According to Christina Stathopoulos on Aporia's website, GenAI "refers to a subset of AI that focuses on producing new and diverse outputs rather than simply responding to inputs. This approach contrasts with conventional AI, where models are trained to replicate existing data patterns."

Common types of GenAI that can be easily found are large language models (LLMs), which are GenAI types trained on a vast amount of data to produce human-like responses to natural language questions with examples like the GPT series from Open AI, Google's LaMDA, and Claude from Anthropic; image generation models, which generate new images based on training data, with examples like DALL-E; and voice generation models, which create new voices also based on trained data, with examples like Replica Studios, Lovo, and Synthesys.

LLMs and SLMs

For application and systems development, the most common GenAI type to be used are LLMs since they generate text-based output that can be used as an input for other software systems. This can be accomplished by calling LLMs through application programming interfaces (APIs) in the form of network endpoints, most commonly accessible through the internet, like OpenAI's GPT API.

LLMs use deep learning to analyze data patterns and generate human-like natural language content. To accomplish this, these models are trained with billions (or even trillions) of parameters and huge datasets of text obtained from books, articles, and websites, among other sources, to produce the required results. Parameters are used to help the LLM decide between different answer choices. Due to this vast amount of training data, these models require a large amount of disk storage and consume considerable processing power.

Since many application contexts do not require this vast amount of pre-training data, SLMs (small language models) can be used on them. These models are trained with less data, which can be millions to a few billion parameters and more specialized text datasets, depending on the intended application. The result is a model that consumes less disk space and processing power but does not generate outputs with the same quality as LLMs and demands training in the specific context where it will be applied.

LLMs Availability and Deployment

LLMs and SLMs are available to be used in different ways. The most common way is to access LLMs that are available on the internet through APIs. For instance, OpenAI offers ways to customize their LLM through custom GPTs that can be trained with data provided by users. This is an option where the LLM and its custom data are maintained by Open AI, where users do not need to worry about the infrastructure. However, the user's custom data can be used to train the LLM in a general context, not only for the user. This default behavior can be disabled by configuration, but it still can be used internally by the company for other purposes. Besides, accessing the LLM through Open AI's API can incur high usage costs, depending on the number of API calls to access the model.

Another way can be by downloading an LLM or SLM from Hugging Face, for example. Hugging Face is a GenAI open-source library site that contains several LLMs and SLMs that can be downloaded to be used on a local server for private and custom usage. The required infrastructure must be managed by the user, though.

LLMs are also being made available by cloud providers, which can be used for private purposes and with custom data. The cloud providers support an ecosystem to be used with the chosen LLM, with options to create data ingestion pipelines with existing technologies, generally specific to the cloud provider. As mentioned before regarding Open AI's example, a cloud provider's infrastructure may also incur high usage costs depending on the chosen infrastructure to host the LLM.

Training LLMs—Reinforcement Learning with Human Feedback

As mentioned before, LLMs are pre-trained in a vast amount of data. To improve the generated output of these models, there is also a fine-tuning phase and a subsequent phase where humans generally play a crucial role. According to Clemens Viernickel on Scale AI's website, they work as part of the reinforcement learning with human feedback (RLHF) technique, which is related to the idea of training "an additional reward model that rates how good a model's response is from the perspective of a human to guide the model's learning process."

Without humans to perform this kind of training, models would not be able to differentiate a good from a bad joke, for example. This RLHF phase is essential to make the existing LLMs generate outputs with the quality that we can experience nowadays.

People who work in this kind of activity perform tasks related to evaluating LLM's responses. They follow strict guidelines about how the work should be done. It generally involves the evaluation of two responses for the same prompt, indicating which response is better and why. The evaluation should be written according to the specified guidelines, and the evaluator must fill in the data as expected. To guarantee the quality of the data, people working as reviewers evaluate if the work has been completed correctly by evaluators. This data is then fed as input for the reward model, which is then used by the LLM to indicate how responses should be generated more acceptably.

Depending on the LLM to be trained, thousands of people may be involved in the training process for several months. Different response types are evaluated for a model. I had recent experience with this kind of job, working as an evaluator of LLM responses regarding software code generation. Other evaluation types may be related to mathematical solving capabilities and LIDAR image analysis, among other types. People are generally paid per hour to accomplish these tasks, whereas the hourly rate depends on the task complexity. As long as the RLHF technique is needed, lots of people will be demanded to work on this kind of task.

LLMs Code Development Usage

It is possible to use LLMs in diverse ways. As mentioned before, calling LLMs through APIs makes it possible for programs to create prompts to send to these LLMs, which reply with outputs that are parsed by these calling programs. This integration possibility is the key to lots of applications that are becoming available.

For example, consider a chess game that needs to be developed. Instead of coding a complete chess AI bot to play with the user, a developer could create prompts specifying the chess piece positions on the board and send this data to an LLM, which would reply with subsequent moves, acting as a chess player. The developer could even specify the difficulty level and style of playing, as well as the output format to be parsed by the calling program. These interactions could then be evaluated, and if they are satisfactory, the chess game could be implemented without much knowledge required by the developer about chess and AI.

LLMs and Traditional AI

Compared to traditional AI, LLMs are accelerating AI usage by systems in general because they are easier to use and integrate with. People who work on software development do not need to have extensive knowledge related to AI to integrate systems; they usually work with LLMs. All they need to know is how to customize and train a particular LLM to be used in the specific context of the system to integrate the LLM, if necessary. On the other hand, traditional AI has advantages when considering performance and predictable output, which are very important factors when compared with LLMs. It all depends on the cost-effectiveness analysis to be done when evaluating the best solution to be implemented.

The Traditional Software Development Process

Software systems are complex. Sometimes it does not seem to be like this because people only interact with the user interface side of them, but most of their size and complexity rely on parts of the system that people in general do not know about. These system parts generally can be located on internet servers that interact with the user's application, which also can have complex software.

To address this complexity, software development activities are mainly composed of the following steps: identify a problem to be solved, plan a software solution to tackle it, craft its architecture, implement the system, test it, deploy it to the intended environment, get feedback to adjust it, and finally deploy it for its intended use. Many people may be needed to accomplish this whole process, depending on the software system size.

From the steps above, the system implementation is the most challenging and time-consuming one. The reason is that when it comes to making the software system a reality, a lot of uncertainties may arise during the development. These uncertainties may be related to unforeseen aspects of the software architecture or unclear requirements, but even when these previous steps are done with enough detail, uncertainty regarding code implementation details is very likely to arise, especially if the development team is not used to the technology involved in those implementation tasks. And even if the team has experience with the related technology, there will always be an effort to develop the related code.

Considering that uncertainty uncovers problems that can affect the system implementation and feasibility, it is very important to uncover and solve these problems as soon as possible. The best way to do this is to start implementing what was planned in the previous steps. It is like compiling a hand-made code: you will only be sure that your code works after the compiler checks that it does not have any compilation errors.

System Maintenance

After a system is implemented and deployed to be used, bug fixes and new features will be demanded for the system to work properly and effectively. While bug fixes are generally minor changes to fix a problem, feature development can be compared to a system development task, but on a minor scale and with the required knowledge about how the system architecture works. These activities demand profound knowledge about how the system works to avoid negative effects on its operation and evolution. They also constitute a major part of the system's life cycle.

How GenAI Can Help in Software Development

GenAI can help mainly with the code implementation process. Since it can generate most of the required code for the system, it can considerably speed up code development and make it faster to uncover technical problems. To accomplish this, an LLM can be fine-tuned and trained to generate the expected code, or it can simply be used through crafted prompts to generate it.

GenAI can also be applied to steps not directly related to coding in several ways. For the architectural phase, it can be used to generate diagrams and documents. For the testing phase, it can help to generate automated tests after those tests are planned based on the related functionality. For the deployment phase, it can be used to help code installation scripts. It can also be used to give ideas of products to address a problem, as well as to leverage common architectural patterns to address some aspect of the system to be implemented. It is like an automated brainstorming process.

Regarding code maintenance, it can be used by developers to help them understand parts of the code that they do not have experience with. It can also be used to help refactor code to adhere to quality requirements and to improve code understanding. Training an LLM or SLM with the existing system's code base can improve the generated code considerably since it will be based on the system's code to generate responses. For example, consider a system that has modules directly based on database tables. With a pre-trained LLM on this system's codebase, prompts like "Generate entity classes using the <system specific> framework based on table <table definition>" could generate code specific to this system's context.

As mentioned earlier, interactions between programs and LLMs are leveraging a vast set of new possibilities with new systems and applications. As an example, nowadays there are code assistants, which are integrated in some IDEs (integrated development environments). They make suggestions for the developer, who can accept them to be applied to the code or not. For instance, GitHub Copilot automatically creates prompts based on the developer's code base to query an LLM, which replies with the desired output that is rendered by the IDE to the developer.

Regarding documentation, it can be automatically generated from the source code by some of those same tools. This is a very important functionality since documentation is an important aspect of a software project, but it is also very difficult to keep it updated with the current code, leading the development team to stop updating and using it due to the huge, required effort.

What GenAI Cannot Help with in Software Development ...Yet!

GenAI is very good at generating text for relatively small tasks, but it cannot be used to generate a complete software system with a couple of prompts. It can be used to generate a small software application, though, to be used as a prototype. Regarding code maintenance, it generally does not perform well for bug fixes. Crafting some prompts to instruct the model to parse the code to fix a bug will not work for bugs that are difficult to find or reproduce. This activity still requires developers with experience in code maintenance related to that system.

Another aspect to observe is that GenAI can be used only to help with software development. The text, images, and code that it produces must always be previously evaluated by developers who know how the system works. This is because the generated code generally does not work correctly or does not address all the prompt's specified requirements, especially if the prompt has complex instructions to be followed. It can also include functionality that was not specified and may also contain insecure code to be used. Simply put, the models available today cannot be trusted blindly for this kind of content generation.

AI and Software Developers

Since AI is rapidly evolving, it is difficult to foresee what exactly will be available in the next few years. The appearance of GenAI technologies leveraged the idea that software developers will not be needed anymore, but the existing technology helps software developers instead of replacing them. AI itself is composed of software that needs developers to create and maintain. Besides AI, the integration activity between existing and new systems with LLMs alone generates a lot of opportunities for developers, who are required to develop this kind of activity, generating new opportunities in this field. Cloud providers are creating solutions integrated with GenAI, which requires people who understand these new features that are becoming available.

Considering this reality, developers are required to understand how to interact with LLMs because they will work with systems that interact with them sooner or later. Regarding code generation, as mentioned previously, the output generated by LLMs still needs developers to evaluate its quality. It is still not possible to develop a real-life working system without developers.

In conclusion, even with part of their work being automated, software developers will continue to be crucial to exploring new challenges and opportunities leveraged by AI and the interaction between human engineers and AI will take the evolution of software development to the next level.

About the Author

Rodrigo Cantú Polo is a Brazilian software development engineer with more than 20 years of experience in different software domains, including corporate systems, Finance, Telecommunications, and Electrical Power systems. He holds degrees in Computer Engineering and Computer Networks and an MSc in Distributed Systems. His current main professional interests are related to backend-related software technologies. He is also a GenAI enthusiast, having worked for Scale AI with reinforcement learning with human feedback (RLHF) to train LLMs.

Email: cantupolo@gmail.com
LinkedIn: https://www.linkedin.com/in/rodrigo-cantu-polo/

EMPOWERING WELL-BEING THROUGH DATA SCIENCE

By Emelie Chandni Jutvik
AI Expert / Engineer in Physics and Mathematics
Norrköping, Sweden

> *Be the change you wish to see in the world.*
> —Mahatma Gandhi

A Journey of Passion and Purpose

I came into this world with an insatiable passion and interest for learning and understanding. My main focus from an early age was to listen and try to understand my physical body. I had a natural passion to explore life and was driven by a strong will to experience the most of my every day. With that being said—I was not searching for endorphins or grand kicks. My adventure consisted of curiously exploring my inner self and what I was feeling in relation to what information I was given from my outer world.

As physical movements and sports were an interest from an early age, it became natural for me to explore the kind of food that made me feel alert and energized. I was eagerly exploring how rest, food, and activity could be combined to best give me as much out of the life that I wanted to live. As I grew older and became a young adult, I soon understood that there was so much more information than movement and food that affected my well-being. All my senses constantly gathered information that I consumed and had to process. As a highly sensitive person, I soon understood that I had to find a more complete approach to achieve my sense of well-being. I had to understand what environments suited me, what kinds of people gave me energy, what kind of job situation was my best fit, how much social activities I could contribute to, and so on. I gradually started to explore every component of my life with the aim to find balance within myself and my life. I hoped to achieve this balanced well-being, so I could experience life as much as I wanted.

I was born in Sweden in the early '80s. At this time, the culture and relation to the physical body were still very unexplored in this part of the world. I was surrounded by an environment that in many ways neglected the natural signals and communication with our human bodies. As a kid with interest in both sports and nutrition I stood out. I was often told to follow the traditional structures and stop thinking outside of the box. I felt like my mindset was met with adversity and resistance, which made me a silent explorer of my inner self for many years. I recall myself daydreaming of a tool that could visualize my inner world, my feelings, and how different life choices affected my body and sense of well-being.

Despite my natural interest to improve my communication and understanding of my physical body, I couldn't always find motivation to keep up with this work. I had been on this internal journey mostly by myself surrounded with environments that neither understand my passion nor my way of living. This had innumerable times made me doubt myself, which caused a lack of motivation to keep going. Many days I wondered how it would be if I could see and show others how small life choices greatly affected a person's well-being in the long term. I always knew I could reach an even deeper understanding and result with myself if I could get some kind of feedback that reinforced my body's signals and helped me to interpret and recognize even the tiniest changes in my body. This would keep me motivated to improve and to make more conscious life choices.

The Unseen Affects and Hidden Knowledge

In 2007, I received my engineering degree in physics and mathematics. I moved to the capital city of Sweden to start a career as a computational engineer. This is the same year I started to understand the value of the inner adventure and exploration that I silently—but passionately—had continued through my life. It only took weeks into my first employment to detect the consequences brought to people that continuously neglected the language of their human bodies. Many of my colleagues became sick due to stress and also were affected by being overweight. This happened despite the fact that many were still in their early thirties to forties. I found it frustrating to see how people who I cared for had created a lifestyle for themselves where their bodies desperately threw communication on them, but they were neither trained to listen nor understand its language. I desperately wanted to simply hand over the sensibility and knowledge that I had gathered during my inner exploration. Again, I found myself frustrated over the fact that there were no existing tools to visualize either my inner well-being or the gathered knowledge that daily guided me to my sense of well-being.

I also found it frustrating that there was no evidence that could support my expressed feeling of well-being. I found it hard to motivate people just by telling them my truth. People often admired my evident health and strength but despised and questioned my life choices. Even though I met many people who wanted to change their way of living, they lacked motivation. Many wanted quick fixes for minimal work effort. Also, many seemed to associate the path to an increased feeling of well-being to a long hard journey that gradually took away all that they defined as fun.

In 2018, I started studying artificial intelligence, which made me see the potential of AI becoming the tool that I had dreamed of for so many years—a tool to reinforce and interpret the human body's language and possibly make humans reconnect and understand their physical systems better. I saw the potential to build a tool that could give us instant feedback regarding a life change—how what you just did could affect you in a positive or negative manner. Further, I could see this tool give predictions—if you keep up with what you just did, you will get this result in the future (positive or negative). This kind of instant feedback and predictions could give us a window into our bodies, which could help us make more conscious choices. We wouldn't have to guess to the same extent—is what I am currently doing working or not? I imagined that an AI-model that predicted a positive future outcome of a given life choice could motivate us to continue pursuing that life choice.

Reconnection Through Data Science

The basic concept of an AI-model (neural network) is that we feed it with experience given by a dataset. The model learns to see trends, relations, and structures in the given dataset and further gives us the possibility to hand out predictions of the future. This means that the AI technology gives us the possibility to feed a neural network with data from our bodies. The network learns to see trends, relations, and structures, which further gives us the opportunity to predict an effect of a specific action or life choice. In the eye of my mind, I see us creating digital twins—trained on each person's individual body data—that continuously gives us an indication of how our life choices affect us.

Now that we understand the basic concepts of the AI model, we could deepen this vision a bit further and look at a practical example. Let's imagine that we had technology that gives us the possibility to continuously measure our blood sugar levels. With this type of technology, we could train a neural network to be able to predict the effect on blood sugar levels depending on what we eat. The network will not know or care about the actual food that we ate. The network will only see and notice a change in blood sugar levels—given by data—and could, therefore, predict how our blood sugar levels will be affected in the near future. This kind of AI-model could be used to automatically regulate insulin levels for diabetics, but it could also be used as an instant feedback tool to learn how different foods affect our bodies in general.

Another example of how to implement neural networks in relation to well-being is to use technology that measures stress levels. Let's imagine an AI-model that is trained to read heart rate variability (HRV). HRV measures the time difference between following heart beats. Research has shown that a low HRV is an indication of high stress levels. With an AI-model trained on HRV we could learn how, for example, different foods, life situations, or people affect our stress levels in the moment. It could also give us an indication of how these different variables will affect us in the near future. I see this kind of AI-model being a tool to identify and visualize different stress factors in a person's life to give them the opportunity to apply new approaches to situations or things that affect them in a negative way.

One of the hardest signals that I struggle to identify is how different supplements impact my body. Is it enough for me to only eat real food or do I have to eat vitamins and minerals as supplements? I would love to have a digital twin that could give me direct feedback on how my body is affected by supplements. I would also be highly interested to see what kind of food has the best impact on my body. I tend to eat a vegan or vegetarian diary since I am naturally drawn to this kind of food. But I would love to be able to get a direct indication of how a diet with more meat would affect my inner world and overall well-being. I would also find it interesting to have a digital twin that could indicate changes that would benefit me if it detected a lack of some vitamin or mineral.

What I discovered during my adventure to find well-being is that this is a fresh product. What makes me feel good and balanced is not a constant. My body's needs are dynamic and to find my sense of well-being I need to listen and flow with my body's communication daily. This is a factor that makes this work so much more complex. There is no quick fix to get to know your body and develop the communication and understanding that is needed to fulfill your sense of well-being. An AI-model in the form of a digital twin could function as a life companion which faithfully follows you through life and gives you predictions and indications regarding given life choices. This life companion allows you less guessing and access to more facts, which could be a great motivator for change and to keep up the work to maintain balance and a sense of well-being.

Technology Rat Race and New Challenges

As you probably noticed, my vision includes technology that is not available yet. The AI-models and the development of hardware technology will always be dependent on each other. We know that the development of hardware and neural networks will always run on parallel development tracks, but none will be a constant leader. Neural networks have been available for many years but the computer technology has not been sufficiently developed. We have, for example, not had computers with enough calculation or storage capacity to be able to use the neural networks to the extent that we can today. In my reasoning and vision described in this chapter, I mention technology that is available in labs but not yet in our everyday life. Technology of this kind—that is more of a general use—is still expensive and is, therefore, inaccessible to most people. From experience we know that the development of technology tends to catch up with the neural networks and vice versa. What we can measure in labs today will most likely be accessible for more common use in the future.

What I find fascinating with this kind of AI-model is that it will require a wider span of corporations to implement. We cannot only focus on the technology related to the AI-model to create this kind of digital twin. We need wearables that can measure our body's in the moment and continuously, we need somewhere to store body data to make it available for the AI model, we need storage space and computational capacity for the algorithm to perform training and predictions, and we need an interface that easily and clearly can present results to a user of this tool. We also need research that can support conclusions drawn by the AI-model. After all, we have to prove that it actually works. Cross functional teams and corporations are vital to develop a prototype of this kind of digital twin. Different companies and areas of knowledge have to come together to develop this tool.

Owning Your Life Path

Not I, nor anybody else, can tell you what well-being is for you. The experienced feeling of well-being is highly individual. I want to create an impartial AI-model that simply visualizes and shows us the effect our life choices have on us. With the feedback and knowledge of our digital twin, we will still be free to make choices based on our individual perception of experienced well-being. The digital twin is firstly thought of as a tool to reinforce the signal from our body, to reconnect and better understand its language. A tool like this could instantly predict and visualize an outcome of a life choice that in reality would take weeks or months for us to see by ourselves. This could work as motivation for us to continue on this new path we have chosen and not give up. Hopefully, a tool like this could help a greater part of the population to find strength and motivation to stop and redirect humanity from continuing to develop diseases caused by neglecting the body's languages.

About the Author

Emelie Chandni is a physicist and mathematician who has worked as an engineer since 2007. During the year of 2018 her interest for AI started to grow through education. Her passion for AI grew even bigger while working on her hobby project: trying to create AI models that can increase well-being in people's everyday lives.

Emelie's creative interest in expressing herself with paper and pen has been a hobby since childhood. Based on her own experiences of psychological and emotional violence, Emelie shares her thoughts and life experiences in her debut book *To Break the Silence* in order to inspire and incite change for the better, both on an individual and a societal level. Emelie has an innate drive to improve the world, which she wishes to do through her writing.

Email: ec.jutvik@gmail.com

CHAPTER 12

A JOURNEY FROM FEAR TO EMPOWERMENT

By Lucy Chen
AI Leader and Advocate
Montreal, Québec, Canada

A ship is always safe at the shore, but that is not what it is built for.
—Albert Einstein

The Dual Nature of AI

In one of the knowledge-sharing sessions I hosted, I posed the question, "What do you think about AI?" Using an anonymous word cloud format, participants' responses began with positive terms like "efficiencies," "optimizations," and "automations." However, a few seconds later, a small but significant word appeared: "fear." As this word grew in the center of the word cloud, it was soon joined by terms like "job replacement," "ethical concerns," "deep fakes," and "privacy violations."

Indeed, AI is like Pandora's box, starting with many beautiful promises but potentially leading to uncontrollable situations such as socioeconomic inequality, market volatility, and even the frightening prospects of weapon automation or self-aware AI. In March 2023, more than 1,100 tech leaders, researchers, and others signed an open letter calling for a six-month moratorium on the development of advanced AI systems. This raises the questions: what is AI bringing to us, and do we know where we are heading?

A Historical Perspective: From Turing to Today

The roots of artificial intelligence stretch back to the 1950s when Alan Turing introduced the concept of the Turing Test. Turing suggested that if a machine could engage in a conversation indistinguishable from a human's, it could be said to be "thinking." This idea sparked the beginning of AI as a field of study, leading to early rule-based systems and basic algorithms.

As computing power grew and more advanced algorithms were developed, AI began to evolve rapidly. The field moved from simple rule-based systems to more complex forms of machine learning, where computers could begin to recognize patterns and make predictions based on data. This shift marked the beginning of AI's transition from a theoretical concept to a practical tool with real-world applications. Early successes, such as the development of expert systems in the 1970s and 1980s, demonstrated AI's potential to assist in fields like medicine, finance, and engineering.

Fast forward to today, and we find ourselves in an era where AI is not only a tool but a driving force behind many of the technological advancements that shape our world. The question that Alan Turing posed over 70 years ago—whether machines can think—remains as relevant as ever, though the context has dramatically changed. But as AI continues to advance, new questions arise. Are these machines truly "thinking," or are they simply executing highly sophisticated algorithms? Is there a point at which AI could develop self-awareness, crossing the line from mere tools to entities with their own consciousness? These questions are no longer purely theoretical.

As we ponder the future, we must ask ourselves: where is AI taking us? Will it continue to be a powerful tool for human advancement, or will it reach a point where it challenges our understanding of intelligence, consciousness, and what it means to be human? The answers to these questions will shape not only the future of technology but the future of society itself.

The Seven Stages of AI Evolutions

AI's evolution can be seen in seven stages. The first stage involved simple rule-based systems and basic machine learning models. I remember learning one of these early models in my university classes. It was in Excel, leveraging a rudimentary system that helped predict and optimize inventory needs for a small retail company, such as forecasting demand for seasonal products, determining reorder points, and minimizing stockouts. Even though it was basic, it saved countless hours of manual calculations on lead time and consumption, and significantly reduced overstock and understock situations.

In the second stage, AI developed the ability to recognize patterns and make increasingly accurate predictions. For example, in one of my past positions, we implemented an AI system to analyze customer data and predict purchasing behaviors, and leverage regressions and clustering models, and we were able to accurately target the right audiences for the marketing campaigns and optimize advertising spending. This system could identify trends and preferences we hadn't noticed, allowing us to tailor our marketing strategies more effectively.

The third stage saw AI tackling complex problem-solving tasks. One of the classic examples is digital twins, which are virtual replicas of physical systems. These digital twins allowed for real-time monitoring and predictive analysis, significantly enhancing decision-making processes. By simulating different scenarios and outcomes, digital twins provided invaluable insights into system performance and potential issues, leading to more efficient and proactive management strategies.

By the fourth stage, AI demonstrated significant advancements in creativity, generating original ideas and content. Recent GPT models and state-of-the-art image generation tools showcased the ability to create visuals that transcend conventional human imagination. These advanced models produced scenes that, while physically impossible, maintained logical coherence, captivating and astonishing viewers with their innovative and surreal representations. This is also the stage where AI became more controversial, such as in debates over the ethical implications of AI-generated deep fakes and concerns about the potential misuse of AI for creating misleading or harmful content.

The fifth stage brought emotional intelligence, enabling AI to understand and respond to human emotions. A notable experiment by researchers at Stanford University and the University of Washington involved creating a highly realistic AI-generated video of former President Barack Obama. The AI system mimicked his voice and facial expressions, generating a video where he appeared to say things he never actually said. This project demonstrated the AI's ability to capture and replicate subtle emotional cues such as tone of voice, facial expressions, and body language, showcasing an advanced level of emotional intelligence.

The sixth stage marked AI's self-awareness, sparking profound philosophical discussions about consciousness. While true self-awareness in AI remains theoretical, experiments with advanced language models like GPT-3 and GPT-4 have engaged researchers in exploring the potential for machines to exhibit signs of understanding and reflect on their existence. These experiments have fueled debates on the nature of consciousness and the ethical implications of developing self-aware AI.

Finally, in the seventh stage, AI seamlessly integrated with human consciousness, creating a symbiotic relationship that expanded the boundaries of human intelligence and potential. This final step led to groundbreaking discoveries, with humans and AI working together to explore new frontiers. By merging human intuition and creativity with AI's computational power and data processing capabilities, this integration ushered in an era of unprecedented innovation and exploration, allowing for advancements in fields ranging from medicine and science to art and philosophy.

Embracing AI: The Journey of Learning and Application

The dynamic interplay between human cognition and AI presents new challenges and opportunities, pushing the limits of what is possible and constantly redefining the frontiers of knowledge and creativity. As AI continues to transform industries and reshape the future, many are eager to dive in but may find the initial steps daunting. The key to beginning this journey is to establish a solid foundation in the basics of AI, starting with an understanding of core programming languages and fundamental concepts. For those already familiar with computer science, languages like C and Python are crucial. These languages serve as the building blocks for many AI models and applications, offering a gateway to understanding how machines learn, process data, and make decisions.

An interesting anecdote from a friend of mine illustrates how perceptions of programming languages have evolved. She's from China and started learning coding in the early 2000s in a college in Canada. On the first day of class, the professor asked, "What language do you know?" She humorously responded, "English and Chinese," thinking of spoken languages rather than programming languages. At the time, this was a humorous misunderstanding, highlighting the gap between human languages and the technical languages used in computing. However, today, with the advent of AI, this distinction is beginning to blur. Modern AI systems, powered by natural language processing, allow users to interact with models in human languages like English, Chinese, and many others. This development has lowered the barrier to entry, enabling more people to engage with AI without needing to master traditional coding languages.

As AI becomes more accessible, the focus shifts from merely mastering technical skills to finding the right balance between human expertise and AI capabilities. While understanding the basics is important, the real challenge lies in effectively collaborating with AI to address real-world problems. This partnership requires harnessing AI's strengths in data processing and pattern recognition, while leveraging human creativity, intuition, and ethical judgment to guide meaningful decisions. The key to unlocking AI's full potential is recognizing where human insight and AI's computational power can complement each other and share success.

Ethical Consideration and Moral Dilemmas of AI

With the lowering barriers to start learning and using AI, we need to truly consider what AI is bringing us and how it impacts our lives. The accessibility of AI tools and education means more people than ever can contribute to and benefit from AI innovations. However, this also raises important questions about ethics, privacy, and the societal implications of AI. As we integrate AI into more aspects of daily life, we must ensure that its development and deployment are guided by principles that prioritize human well-being, fairness, and transparency. Recently, I read an article about AI and empathy, an emotion traditionally thought to be uniquely human. A Google research paper, "Towards Conversational

Diagnostic AI," reveals that a healthcare-specific AI system outperformed or matched human doctors in simulated patient interviews and diagnosing based on medical history. This leads us to question whether AI will replace jobs we consider essential and irreplaceable.

Consider the implications of AI making decisions for us. For example, with self-driving cars, we entrust our lives to AI models. MIT's moral machine platform explored human perspectives on moral decisions made by AI, presenting dilemmas like choosing between two evils when a self-driving car's brakes fail. These scenarios challenge our moral standards, as we must decide who is responsible for life-and-death decisions.

Another well-known ethical dilemma is the "trolley problem," where one must choose whether to divert a runaway trolley to save five lives at the cost of one. This scenario forces us to confront difficult questions about the value of life and the morality of decision-making in life-and-death situations. As AI systems increasingly take on roles that involve such critical decision-making, the ethical complexities become even more pronounced. If an AI is programmed to make decisions in scenarios like the trolley problem, it raises significant concerns about accountability and responsibility. So, who is ultimately responsible for the outcomes—those who programmed the AI, the operators who deploy it, or the AI itself? These questions challenge our current ethical frameworks and demand a careful consideration of how we design, implement, and oversee AI systems, ensuring that they align with societal values and moral principles.

Elon Musk, when creating SpaceX, cited the Foundation series and Zeroth Law as fundamental, inspired by Isaac Asimov's novels. The Zeroth Law states that a robot may not harm humanity or, by inaction, allow humanity to come to harm. This principle underscores a profound commitment to the greater good, guiding the ethical development and deployment of technology. By adhering to such principles, we can ensure that advancements in AI and other technologies serve to enhance human welfare, address global challenges, and promote a future where technological progress aligns with the values of safety, equity, and collective benefit. As we continue to integrate AI into various facets of life, it is crucial to embed these ethical considerations into our development practices to safeguard against potential harms and drive positive societal impact.

Overcoming the Fears and Practical Strategies for AI Integration

Returning to the beginning of the chapter, it's not surprising that some survey respondents expressed fear of competing with AI or even using AI tools. However, if we make friends with AI, it can be the fuel that propels our work, allowing us to focus on roles that only humans can fulfill. Gen Z's rapid adoption of AI compared to Gen X illustrates this point. With Gen Z entering the workforce, their comfort with AI may lead to faster technological integration, boosting productivity but also posing intergenerational collaboration challenges. Once we unleash the power of AI, there's no turning back, making it crucial to upskill the existing workforce and foster an AI-friendly culture.

To reduce fear of AI and foster a more positive outlook on its potential, I suggest three stages for approaching this evolving field: integrating AI into your life, bringing AI into your work, and staying up to date with AI developments. As someone who has embraced AI, I've found it an incredible ally. AI-powered virtual assistants manage my schedule, navigation apps optimize my commute, and machine learning algorithms in my fitness app personalize my workout routines. These tools have made my life more efficient and enjoyable, demonstrating AI's potential as a helpful companion.

Second, in your professional life, the applications of AI are profoundly transformative across various dimensions. For instance, in supply chain management, AI can revolutionize inventory management and demand forecasting. By analyzing historical data and market trends, AI algorithms can predict future demand with remarkable accuracy, allowing businesses to optimize stock levels, reduce excess inventory, and minimize stockouts. This predictive capability ensures that resources are allocated more efficiently and can significantly cut costs associated with overproduction or shortages.

In the marketing realm, AI enhances customer segmentation and personalization. Machine learning models analyze consumer behavior, preferences, and purchasing patterns to create highly targeted marketing strategies. This not only improves customer engagement and conversion rates but also enables more effective allocation of marketing resources, maximizing ROI.

Furthermore, AI's role in reducing CO_2 footprints is increasingly critical. For example, AI-driven energy management systems optimize energy use in manufacturing processes and buildings, identifying opportunities for reduction in energy consumption and carbon emissions. Predictive maintenance powered by AI helps in scheduling timely repairs and avoiding energy waste due to equipment malfunctions. Additionally, AI can assist in developing and managing sustainable supply chains by optimizing routes for transportation, thereby reducing fuel consumption and emissions.

Lastly, actively learning about AI is essential. Embrace the opportunity for growth and innovation, even if you're a senior executive. Attend AI workshops, engage with experts, and experiment with AI tools. By demonstrating a willingness to learn and evolve, you inspire others to do the same. AI is here to augment human intelligence, not replace it. By approaching AI with an open mind, focusing on its potential to solve problems and create value, we can transform fear into excitement about the possibilities AI brings to our industries and beyond.

In conclusion, our journey through the evolution of AI underscores a profound truth: AI is not our adversary but our ally. This lies at the foundation of the AI Mindset. As we traverse from the initial fears of job displacement and ethical dilemmas to the transformative power of AI-driven innovation, it's clear that collaboration between humans and AI holds the key to unlocking unprecedented potential. Just as quoted from the beginning, "A ship is always safe at the shore, but that is not what it is built for." Einstein's metaphorical ship is designed for exploration beyond the safety of the shore, so too is AI designed to propel us beyond our current limitations.

Embracing AI's capabilities while maintaining our unique human attributes—creativity, intuition, and ethical judgment—enables us to navigate the complexities and opportunities that lie ahead. AI, when seen as a collaborator rather than a competitor, can enhance our abilities and drive innovation, helping us solve pressing challenges and create new possibilities. By overcoming apprehensions and integrating AI into our lives and work, we unlock its full potential to complement and amplify human efforts. This collaboration allows us to focus on roles and tasks that only humans can fulfill, fostering a productive synergy that propels us toward a brighter future. As we continue to adapt and learn, AI will serve as a powerful ally, reinforcing that together, we can achieve extraordinary advancements and address global challenges with greater efficacy and insight.

About the Author

Lucy Chen was born in China and raised in Canada. After graduating with a finance degree, she began her career as an analyst but soon discovered a passion for data science, leading her to pivot into that field. Following a few years in the industry, Lucy pursued a master's degree in applied Analytics from Columbia University. She then led a research group at a global organization specializing in innovative data-driven projects, utilizing machine learning, artificial intelligence, and predictive modeling to enhance business outcomes and optimize customer experiences. Lucy is also committed to advancing the energy transition and reducing carbon footprints through data-driven solutions. In addition to her professional work, Lucy is a passionate advocate for AI education. She serves as a learning facilitator for 2U, the company behind edX, and facilitates MIT Sloan & CSAIL Executive Education programs including Machine Learning in Business and Artificial Intelligence Implications for Business Strategy.

LinkedIn: https://www.linkedin.com/in/lucy-chen-data-science/

CHAPTER 13

SECURITY FOR ARTIFICIAL INTELLIGENCE

By Hermann Escher
Co-founder of A + E Informatik Security, AI Expert
Zurich, Switzerland

In Greek mythology, Pandora opened a box that unleashed all the evils of the world, leaving only hope behind.

This old story serves as an apt metaphor for the challenges and risks that come with developing and implementing artificial intelligence (AI). While AI has the potential to provide immense benefits and solve numerous problems, it also carries the risk of unleashing unpredictable and harmful consequences.

Pandora's box serves as a reminder that while curiosity and progress drive innovation, they must be balanced with responsibility and caution, especially in the context of AI. In the world of AI development, this means that we must not only celebrate the technological possibilities, but also carefully consider the ethical, legal, and safety aspects. This is the only way we can ensure that hope—the positive effects of AI—prevail and that the evils are kept at bay.

The possibilities and developments in the AI field are not yet fore-seeable. The use of the various types and methods of AI is already a benefit for people and companies. The question of whether AI will have a negative impact is increasingly emerging. In many cases, however, the positive view prevails. A problem is that with people's enthusiasm for AI solutions, people often underestimate or forget the need to be cautious and consider the security of the AI model. According to experts, no AI service can guarantee complete security. People are simply expected to believe that AI models are secure. When we consider what types of attacks on AI systems are possible, it's apparent we must proceed with the utmost caution when using AI systems.

Neural networks, inspired by the structure of the human brain, are crucial components of AI systems. They are able to process large amounts of data and adapt themselves. However, these models are susceptible to manipulation, which emphasizes the need for security measures. Manipulation of AI models can have serious consequences, for example, if they are compromised in security systems or promote industrial espionage. To ensure the integrity of AI systems, compre-hensive security standards, continuous monitoring, transparency, and advanced detection methods are crucial. In addition, the standardization and certification of AI protection and security is essential to ensure the general trust and safe use of AI. It is important that the security of AI systems is continuously improved to minimize potential risks and take full advantage of AI technology.

While I believe that AI has the potential to improve people's experi-ence both at work and in life, I think part of the AI Mindset that people develop should be caution. In this chapter, I attempt to highlight many, but not all, potential risks to using AI.

Excessive Trust

Excessive reliance on AI carries significant risks and requires a balanced approach to this technology. The reasons for over-reliance on AI lie in a fascination with technology, a lack of understanding of how AI works, an overestimation of AI capabilities and efficiency, and blind trust in authority. It's important to consider the potential risks, including securi-ty threats, ethical concerns, loss of human capabilities, reinforcing bias, economic risks, privacy issues, societal impact, and lack of transparency. Responsible use of AI requires careful consideration of opportunities and risks, as well as clear regulation to minimize the negative impacts. Companies and governments should actively address these challenges while taking advantage of AI technology to develop innovative solutions.

To strengthen trust in AI systems, comprehensive educational measures and the promotion of critical thinking are necessary. Government regulation, diversity in AI development, regular reviews, and ethics play a central role in ensuring a balance between humans and machines. It is important to maintain a healthy level of skepticism about new technologies in order to minimize the risks and reap the benefits responsibly.

Excessive Incapacity to Act

The growing autonomy of AI requires careful ethical consideration and clear guidelines. It is crucial to ensure fairness, transparency, and accountability along with the aim of minimising risks. Multidisciplinary collaboration and flexible regulations are necessary to ensure the responsible use of AI. Ultimately, it is crucial to balance innovation and protection and increase trust in technology.

The challenge is to use AI in a targeted way to expand our capabilities without becoming completely dependent on it. It is up to us to use the potential of AI while always keeping an eye on the ethical and social implications.

Excessive Dependence

In a digital world where AI systems are becoming increasingly important, it is important to observe appropriate security measures. The excessive use of AI carries risks that should be minimized through enhanced security measures such as encryption, access controls, and training. Compliance with data protection regulations and ethical principles is crucial to promote trust in AI systems and prevent data breaches. It is important to understand potential risks and take appropriate measures to increase trust in AI technology.

For developers of large language models, it is crucial to carefully minimize risks and ensure the confidentiality of sensitive information. This requires measures such as data cleansing, anonymization, and regular security checks. Proper security precautions, including robust encryption and infrastructure security, are essential to prevent serious data leaks. A holistic security strategy and data protection are essential to maintain trust in AI technologies and protect privacy. Technical and organizational measures such as neuromorphic encryption, differential privacy, and regular security audits play an important role in this. Particular attention must also be paid to the processing of sensitive data and ethical behaviour.

Shadow AI in Companies

Shadow IT, including AI systems, refers to tools and systems used in organizations without official approval. Such systems can be purchased and used by individual departments or employees, often without the knowledge or consent of those responsible.

Although shadow IT can increase efficiency and productivity by allowing users to bypass the limitations of official IT systems, it comes with risks of possible data leakage.

Often, such tools and systems do not meet the organization's security standards and protocols. They lead to problems because generated data is processed and stored outside official systems. This makes it impossible to have an overall view of the organization's data integrity and security. At the same time, there is a risk that data and information will flow uncontrollably and possibly violate applicable guidelines and laws.

Therefore, it is crucial that organizations implement strategies to identify shadow IT. This includes conducting regular IT audits, introducing guidelines for the use, approval, and prohibition of IT systems, tools, and applications, as well as training and raising awareness among users about the risks and responsibilities.

The Snowball Effect

The snowball effect with manipulated AI tools refers to the process where the use of these tools increases rapidly over time. The more data is collected and the AI learns and adapts, the greater the results or impact of using AI can be, like a snowball that gets bigger and bigger as it rolls downhill. The snowball effect is driving the development of AI as self-learning algorithms are constantly improving themselves. Networking and scalability spread innovations through duplication, so technological advances enable more complex tasks and interdisciplinary collaboration creates new applications. The dynamics in the snowball effect lead to faster innovations, increased efficiency in companies, advances in research and medicine, and broader access to knowledge. However, the effect can also be used to build false data or information into an origin system through an attack. It is important to manage risks such as data misuse and to adopt a balanced approach that combines technological, ethical, and legal aspects in order to reap the benefits of AI and minimise risks

Data Manipulation

Data manipulation refers to the process of adapting, modifying, or controlling data to achieve a specific goal. This can be done in a variety of ways, such as cleaning, transforming, and merging records. Attacks on AI systems pose a major threat to the integrity and reliability of AI systems. The variety of attack methods requires a comprehensive understanding of AI protection, security, and versatile defense strategies. Risks can range from data poisoning to integrity attacks, potentially causing effects ranging from performance degradation to serious security and privacy breaches. Effective countermeasures include technical solutions such as stable model architectures and continuous data validation, but also organizational and ethical approaches.

Manipulation of Training Data

The security of AI systems is of very high importance due to their susceptibility to manipulation by training data, complex models, and generalization capability. Attackers exploit vulnerabilities in publicly available algorithms and execute various attacks such as evasion, data poisoning, and Trojan attacks to undermine the integrity and reliability of AI models.

To protect AI systems, technical solutions such as adversarial training and anomaly detection are required, as well as organizational measures such as security audits and employee training. Regulatory and ethical considerations also play a role, as does the development of industry standards and legal frameworks. The training data for a model must be validated, anonymized, and impossible to change. Applicable procedures still need to be developed for this purpose.

Insider Theft of AI Models

AI models can be stolen in a variety of ways, including insider theft, cyberattacks, reverse engineering, model extraction, and side-channel attacks. Insiders have direct access while hackers break into systems. Reverse engineering and model extraction are techniques for recreating model structures. Data poisoning damages the model by manipulating the training data, and side-channel attacks use side effects to gain information. A well-known example is the controversy over stolen AI models, as in the Tesla AutoPilot case in 2019.

The scandal, in which former Tesla employees were accused of leaking confidential AutoPilot data and algorithms to Chinese competitors, has highlighted the vulnerable security of advanced AI systems in the automotive industry. This led to a broad debate about security measures and the protection of AI technologies in the automotive industry. In response to this incident, many automakers have strengthened their security protocols and invested in protecting their AI technologies.

Referral Attack

Referral attacks aim to find out whether certain data has been used in the training of AI models. Attackers analyze model outputs to draw conclusions about training data, which puts the security of AI systems at risk. Overfitting facilitates such attacks, as models then react differently to training data than to new data. Shadow models, for example, can be used by attackers to mimic the behavior of the target model and perform more precise attacks.

To protect against these attacks, measures such as differential privacy, regularization techniques, model compression, federated learning, strict access controls, and data minimization can be used to make it more difficult to distinguish data points and minimize the risk of data leakage.

AI Supply Chain Manipulation

AI models, and the platforms they run on, are critical components of the AI supply chain, which encompasses everything from the development and deployment environments to the tools and libraries used. For example, a recently discovered vulnerability in Hugging Face, a popular machine learning platform, highlights how weaknesses in these components can allow attackers to gain unauthorized access to sensitive data or manipulate models. Such vulnerabilities within the AI supply chain can compromise the integrity and security of downstream applications, emphasizing the need for robust protection at every stage of AI development and deployment.

The AI supply chain, which is essential for the development and operation of AI systems, is an attractive target for cybercriminals. It is a complex ecosystem that has various attack vectors and vulnerabilities. Attacks can occur on different levels of the AI supply chain, including:

- The compromise of development tools and environments
- The injection of malicious code into open source libraries
- The manipulation of software updates and patches

These attacks can lead to vulnerabilities being built into AI systems that can be exploited later.

Manipulation of Transfer Learning

The concept of transfer learning is crucial in the field of machine learning, as it allows pre-trained models to be adapted to new tasks. It starts with training on an extensive dataset, followed by fine-tuning for more specific tasks. The method saves data and computational time and often improves the accuracy of the models, especially with limited amounts of data. The areas of application are diverse—from image recognition to medical diagnostics. Challenges such as the negative transfer phenomenon require the development of strategies to avoid performance impairments.

When learning transfers, both the benefits and the security risks must be taken into account. Adversarial attacks can fool models while source errors go undetected. The complexity and risk of model hijacking make it difficult to track decisions. It is important to minimize these risks and ensure the robustness of the models through measures such as adversarial training and improved validation. Ethical guidelines and continuous learning are critical to adapting models and improving their performance. Transfer learning offers promising developments such as zero-shot learning and multimodal learning, and adaptive mechanisms and data protection techniques could play a significant role in the future.

Model Denial of Service

Attacks on AI models by model denial of service (MDoS) are different from traditional denial of service (DoS) attacks. They exploit vulnerabilities and resource requirements, overload processing capacity, and can cause performance degradation by manipulating the input format. Protecting AI systems requires comprehensive security measures, including the implementation of security protocols, firewalls, regular security reviews, and training for employees. This is crucial to ensure the integrity and performance of AI systems.

Unsecure AI-Plugins

Secure plugin design is crucial for the security of AI systems. The modular architecture offers flexibility, but it also comes with risks, especially with regard to the Open Worldwide Application Security Project (OWASP) top 10 security risks. A strong focus on security in the development of AI plugins, including secure data transfer and access control, is essential. Regular checking and updating of the plugins is necessary to fix security vulnerabilities. A holistic security strategy is required to ensure the integrity and confidentiality of AI systems.

The recently discovered vulnerabilities in AI plugins highlight the risks of insecure designs. For example, a vulnerability was discovered in ChatGPT that allowed malicious plugins to collect sensitive information and insert malicious content. There is an urgent need to improve security standards to implement strong authentication mechanisms, data validation, secure communication, and regular security checks. Version control, isolation, logging, and data encryption also play a crucial role in preventing security risks in AI plugins.

Manipulating AI Data Packets

The importance of AI data packets for the structuring and processing of information in AI systems is crucial. Training data, input data, configuration data, and intermediate results play an important role and must be handled carefully to maximize the performance and efficiency of AI systems.

Ensuring the integrity of AI data packets requires preventive measures such as a solid data infrastructure, regular backups of unmanipulated models, data cleansing algorithms, and employee training. Protections include robust storage and transmission systems, automatic error detection, encryption technologies, continuous monitoring, and fault-tolerant models.

Direct Prompt Injection

Direct prompt injection is a method that enables the manipulation of AI language models through specially formulated text sequences. This poses potential security risks, as attacking parties can send unwanted commands. It is, therefore, crucial that developers and system administrators take proactive security measures, such as implementing input validations, monitoring system activity, and regularly updating security protocols, to effectively contain the potential threat.

To ensure the security of AI systems, robust input validations, AI firewalls such as WAF and security filters, training for developers and users, sandbox environments, prompt engineering techniques, and regular security audits must be implemented. In addition, strengthening security measures through multi-factor authentication and anomaly detection is crucial. Research advances, such as self-monitoring AI systems, adversarial training, and explainable artificial intelligence (XAI), contribute to the safety of AI models and are supported by industry-wide standards to increase confidence in their applications.

Indirect Prompt Injection

Indirect prompt injection is a method that exploits the complexity and flexibility of modern AI models to manipulate system behavior without interacting directly with the user. This is done by deliberately manipulating input data to achieve undesirable results. This can happen through context manipulation, prompt engineering, and data contamination, and poses a serious threat to the integrity and security of AI systems.

Robust input validations, sandboxing, training with manipulated data, monitoring systems, and multi-layered security architectures should be used to defend against such attacks. In addition, regular security audits, strict access controls, transparency, ongoing training, and contingency plans are essential for developers and users of AI systems.

AI Model Inversion Attacks

AI model inversion attacks are highly complex processes in which an attacker reproduces the structure and training data of a machine learning model in order to obtain sensitive information. This requires an in-depth understanding of how the model works and the data, and highlights the need for robust machine learning security measures.

Manipulated automated systems, such as trading systems and scoring models, can cause significant economic damage and endanger public safety, e.g., in the case of autonomous vehicles or medical diagnostic systems. Strict security standards and control mechanisms are essential for the development and use of such systems.

In order to ensure the security of AI systems, technical and organizational measures are crucial. This includes robust model architectures, security audits, and continuous monitoring. Advances in self-learning safety and explainable AI as well as interdisciplinary collaboration play an important role. However, there is still a need for research on ethical aspects and faster methods of detecting attacks.

Output Integrity Attacks

Output integrity attacks aim to falsify AI's outputs without it being immediately noticeable. They can be done by manipulating the input data or influencing the AI model or external factors. The impact can be subtle, but it has far-reaching consequences, as it can undermine trust in AI systems and provide financial, political, or personal benefits to attackers.

Concluding Thoughts

It is our responsibility to find a way to reap the benefits of AI without sacrificing our independence and critical capabilities. By prioritizing education, ethical frameworks, responsible development, and mindful AI usage, we can cultivate an "AI Mindset" that ensures harmonious collaboration between humans and machines.

About the Author

Hermann Escher is co-managing director of an IT company and head of the security and artificial intelligence departments. He works on strategies for various AI solutions that improve and simplify the work. The same applies to its safety and protection. With 35 years of expertise in digitization and IT security, Hermann brings a wealth of experience to advancing AI safety.

This year, Hermann successfully completed the course in Artificial Intelligence in the Business Environment at the renowned Massachusetts Institute of Technology (MIT). The training makes it possible to use AI technologies in a targeted manner in entrepreneurial contexts. With these competencies, he plans to support companies in the optimal use of AI-based solutions.

Website: https://www.aeinformatik.ch/
LinkedIn: https://www.linkedin.com/in/hermann-escher-23b856a/

CHAPTER 14

NAVIGATING DIGITAL TRANSFORMATION IN REGULATED INDUSTRIES: LEADERSHIP AND INNOVATION

By Isabelle Flückiger
Board Member for Digital Transformation, AI Advisor
Zurich, Switzerland

Whether you think you can,
or you think you can't—you're right.
—Henry Ford

"No, we decided not to make any additional investments. We must be compliant, nothing more." The words echoed in my mind, a stark reminder of the uphill battle I faced, once again, at another bank. My proposal, a meticulously crafted plan to generate sustainable profits with a minimal additional budget by leveraging advanced machine learning algorithms, was dismissed without a second thought. The potential returns were undeniable—multiple times the investment, recouped in less than two years. Yet, the banks, ever notorious for prioritizing their profits over customer value, made their decisions.

It wasn't the quality of my proposal or regulatory constraints that led to its rejection. It was a deeper, more systemic issue. Banks and insurance firms, conspicuously absent from the lists of the most innovative companies, were entrenched in their ways. They view digital transformation merely as a cost-saving tool to replace humans one-to-one with machines. This narrow perspective failed to grasp the transformative potential of technology. Regulated industries struggle with transformations while constantly seeking salvation in technology hype, such as generative AI at the moment.

After over two decades of evaluating, consulting, and implementing technology-driven innovation in regulated industries, I had accumulated enough traumatic experiences to seek refuge in a well-paid, quiet corporate job. The relentless resistance to change may have worn me down. Tragically, the patterns are always the same, whether before or after the last financial crisis, whether before or after the pandemic or in the current artificial intelligence hype. To this day, I have not figured out what keeps the mindset of the regulated industries so captive, but I can rule out the technology itself. Every time banks are addicted to the latest technology hype.

People often ask why, as a technology and especially an AI expert, I focus so much on mindsets, culture, languages, economics, and behavior. The answer is simple: digital transformation is not just about technology. It's a way of thinking, a mindset, and a way of behaving. Until these industries understand that, true innovation will remain out of reach.

The Challenge Beneath the Surface

I could have easily checked a box, completed the bank's ongoing compliance project, and moved on to the next project. But that would have been too simple, too superficial. Recently, I had an enjoyable dinner with the former head of Insurance and Pensions at the European Commission. One of his "crimes" was introducing a ban on gender-based insurance pricing at the European level. The insurance industry's outrage was so immense you might have thought the entire sector was on the brink of collapse, but finally, nothing happened. He imposed additional regulations on the industry to safeguard policyholders, and the industry's outcry over the perceived futility, the false sense of security it suggested, and the associated costs were overwhelming, and they claimed that it would mean the death of the European insurance industry. Yet, predictably, nothing happened.

During our dinner, he shared, "Regulation is often perceived as a burden, but I aim to stimulate innovation for the benefit of our society beyond any technology, regulation, and laws." His words resonated deeply with me, highlighting the disconnect between the perception of progress and the status quo. The saying "the status quo is the greatest enemy of change" seems to fit perfectly here. But why doesn't somebody in the financial services industry think that no change is the biggest enemy of generating revenue and growth? The not-so-obvious and noticed aspect is the culture of marginal cost of companies whose core operational model revolves around money and financing.

Additional costs are always seen as a burden, diminishing the margins of an existing business. This marginal cost mindset is a strategy of fear aimed at protecting a current business. It assumes that the business will still exist in a few years without considering the possibility that this strategy might make the business obsolete. Marginal costs are rarely low. All the global banks and insurers that collapsed or had to be rescued failed because of this mindset. They couldn't anticipate the true total costs of their marginal thinking. When the status quo core business disappears, marginal thinking will become a full-cost accounting for costs, markets, and reputation. Trapped in this mindset, innovation makes no sense from their perspective, and all rational and logical arguments are of no help. And I know the financial services industry as a very engaging expert in that mindset. My proposal to leverage advanced machine learning algorithms in the compliance project to increase revenues was rejected due to this marginal cost thinking.

But I wanted to always dig deeper. What could change this marginal thinking and turn it around? How should company leaders think and act to be the greatest enemy of marginal thinking and the status quo?

The Three Disciplines of Successful Navigators

While my frustrations with the financial industry were many, my work in the life sciences sector often provided a refreshing contrast. In this field, placing bets is part of the business. With 90% of all developed clinical drugs never making it to market and those costing over a billion dollars, there is no room for marginal thinking. So, what do companies do in its absence? A closer evaluation of the market leaders' approaches revealed three focal points for successfully implementing digital transformation: strategy, measurement, and mindset.

Regulated companies often mistake economic and regulatory requirements for strategy, adding some technological elements and calling the result digital transformation. But is that truly a strategy? The best litmus test for a valid strategy is whether a counter-strategy exists. Applying this test to a strategy of "compliance" would suggest that "non-compliance" is a valid alternative. As many bank failures have shown, it is not. So, compliance is not a strategy. But would compliance driven by technology be a valid strategy? Is non-technology-driven compliance even possible in today's world? Hardly.

One life sciences company I worked with added "data" to its strategy. The deeper meaning was that employees should treat data and data-driven decisions rigorously as they do drug research and development, ensuring that data benefits patients under the same ethical and regulatory standards. As a by-product, data would also boost product launch success and enhance the company's overall value. This is not only a great strategy but a prime example of using regulation as a catalyst for digital transformation and innovation. This strategy balances innovation and regulation while integrating patient, employee, and investor perspectives. Most notably, it has no direct reference to technology. This strategy focuses on a change in thinking and embodies the mantra of sustainability over hype. It shows that transformation needs a strategy that enables a shift in thinking and must always be aligned with strategy.

A second essential point is measurement. Peter Drucker's often-quoted statement, "If you can't measure it, you can't improve it," is particularly true in digital transformation. This brings me to another misconception I often encounter, one that frequently faces even greater resistance. The 80-20 principle, suggesting that 80% of the total result can be achieved with 20% of the effort, is as sure as the amen in church. While this sounds tempting, it is not. In my experience, this 80% too often focuses solely on technical functionalities. These 80% are not contextualized, meaning they are not measured against any standard. Against what should 80% be achieved?

Moreover, 80% is useless if customers expect 96% or if 89% is needed to remain competitive in the market. This issue is prevalent across practically every key performance indicator (KPI) I have encountered. Many companies seem to have hundreds of KPIs, losing sight of what is essential and failing to consider the context in which they should be measured. The most successful companies I have worked with manage their business with just four KPIs, each with a clearly defined context. As seen in the life sciences company's strategy example of "data," these KPIs reflect the perspectives of the company, customer, employee, and investor. This is the high art of leading a company.

The type of KPI also plays a crucial role. Successful companies focus on metric-based KPIs instead of task-based or, even worse, date-based KPIs. It does not matter if and when a task is completed. Metrics-based KPIs measure progress over a period towards a strategic goal, allowing for influence and adjustment. This means that change is monitored and measured—something fluid—and you interact and influence the path to the goal. This reflects a liquid mindset in a fluid environment.

The two aspects of strategy and measurement naturally lead to the third element: mindset. Analyzing mindset often left me in a state of confusion, primarily due to my initial narrow thinking. In attempting to distill the successful navigator's mentality into a few streamlined characteristics, I repeatedly failed, feeling disoriented and enveloped by a dense fog. It was only when I realized that there is no single mindset, but rather that the decisive characteristic of a successful mindset is its liquidity, that I understood the difference. Mindset has many facets, evolves over time, and adapts to changing circumstances. It is the only way to keep pace with, or even outstrip, the ever-accelerating regulations. In this way, regulation is not just a necessary burden but an enabler to make business and benefit customers. It facilitates the business of the future and prevents a system of preservation. It changes how companies operate, communicate, and interact with their customers, employees, and shareholders. This fluid steering is digital transformation at its core.

Most companies react to increasingly more regulations at an accelerated pace by launching ever larger and faster projects. The main problem lies in the word itself: they react. It's like a horse race where banks keep buying faster horses yet still lag behind the regulations. This approach vividly illustrates the marginal cost mindset. If they conducted a full-cost analysis and realized their obsolete business assumptions, they would stop buying horses and switch directly to a car. Another consequence of this marginal cost thinking is the "think big" trap in transformations and the tendency to think in only one direction—big. I've seen countless projects over my 20 years of experience that went off track and failed. By the time the failure was realized, millions of dollars had already been wasted. Companies look to market leaders and want to catch up to them in one big project, regardless of how often those market leaders failed, learned, and tried again. One embodiment of this is the attempt to build a jack-of-all-trades solution that solves all problems, typically distancing from the core business. Instead of "think big," successful implementers have big visions and then craft scenarios, analyze them, and place their many bets accordingly. This means they adopt a full-cost approach, fully aware that they must be prepared to rethink and realign their business.

What is often overlooked is that regulations are driven not only by regulators but also by investors and the capital market. The current sustainability regulations are such an example. While everyone talks about the official government-driven standards, which mainly apply to large corporations, they neglect the fact that investors expect the same from firms that officially would not have to follow these requirements. Without this, bonds, for example, can no longer be issued and placed in the capital market.

Regulation becomes an unavoidable and compelling driver of growth and value generation regardless of whether you must formally follow the regulations. This, in turn, requires again a new mindset. Leaders are, therefore, forced to maintain and increase shareholder value while managing the skyrocketing costs of regulations. Consequently, shareholder value and regulatory costs can no longer be separated. This isn't just about cost accounting; it requires unified thinking. All of this shows that the mindset is contextual and considers the varying aspects of time. Processes in regulated areas are usually complex and lengthy, requiring much diligence. This complexity prevents rapid movement. So, the liquid mindset enables leaders to act agilely in various dimensions, from big to small project sizes and fast to slow implementations. It's about being adaptable and fluid, ready to pivot and realign as needed. Ultimately, it's not just about keeping up with change—it's about leading it.

The New Leadership Paradigm of a Liquid Mindset

Digital transformation and AI in regulated industries requires a new type of leadership to embrace, drive, and nurture innovation. This leadership metamorphoses into a multifaceted role—navigator, influencer, enabler, leader, decision maker, and value driver. Company executives are usually not chosen for these attitudes nor chosen to bring cultural changes in this mindset, and those who did were often replaced by minds focused on hard shareholder-driven thinking. But these qualities are not mutually exclusive. As a board member of an energy company, I experienced this firsthand during the recent energy crisis in Europe. Many companies plunged into crisis and even required government intervention to survive, but we managed to emerge from the stormy seas even stronger than before. While strategically and financially managing a regulated company during a crisis, our game changer was adopting a liquid mindset in this fluid environment, where there are no constants and no end—neither in regulations, market competition, or technology, which evolves at near-light speed.

The same patterns were observed in the life science industry during the pandemic. So, what we can learn from the leaders in regulated industries is that a radically embracing mindset forms the foundation for success. They not only accept the polarity of actions and decisions but consistently seek change. When I work with these match winners, their complementary mindset is ever-present. They understand that there will never be an end; it's fluid, nothing is stable. It's about un-hyping the hype, focusing always on sustainable business and growth. Leaders need consistency, and consistency is a journey that requires embracing change and anticipation. With a liquid mindset in today's fluid environment, a leader is able to handle the polarity of digital transformation and AI in regulated industries. They become the true navigators of change, steering their companies not just to survive but to thrive in an ever-evolving landscape.

About the Author

Isabelle Flückiger enables transformational thinking for organizations in regulated industries for over 20 years. She earned a PhD in mathematics from ETH Zurich, focusing on algorithms now essential in AI. Isabelle combines technical expertise with a deep understanding of human and

131

behavioral aspects, which she believes are crucial to project success and cultural change.

In her consulting career, she has built several new service fields and teams, maintaining a low turnover rate of less than 5%. Her integrated approach has gained her global recognition. As a partner and managing director at Accenture, she led the applied intelligence business for the financial services industry in German-speaking Europe.

Isabelle has also been a trusted advisor to various regulators. Currently, she serves on the boards of companies in regulated industries, focusing on digital transformation, technology, strategy, and innovation. She is a part-time lecturer at ETH Zurich and an expert for Innosuisse, the Swiss government's innovation agency. Isabelle works as an independent consultant, speaker, and change manager.

Isabelle's mission is to help organizations stay future-relevant by fostering a proactive, adaptable mindset. Isabelle's career exemplifies integrating technical skills with a focus on human factors and driving sustainable change.

Email: isabelle.flueckiger@pm.me
Website: www.isabelleflueckiger.com

BUILDING AN AI ROADMAP FOR YOUR ORGANIZATION—A FIVE-STEP APPROACH BASED ON VALUE

By Carl Jones, MBA
AI & Business Value Consultant
Melbourne, Victoria, Australia

It is not from the benevolence of the butcher, the brewer, or the baker, that we expect our dinner, but from their regard to their own interest.
—Adam Smith, 18th Century Pioneer of Economics

Imagine the horror! Your somewhat unreasonable boss Graeme has been reading a lot about artificial intelligence, and particularly generative AIs like ChatGPT. One afternoon he comes to you and says "Hey, this AI stuff is incredible, it is so cool, let's see how we can use it in our organization. Investigate this AI thing and present our options back to the senior management team in seven days please." Oh no, where would you start?

This is not actually an extreme example. During my time working in artificial intelligence with corporate, government, and not-for-profit organizations, I've seen this scenario many times. It's becoming common. Executives often see the stories of potential benefits, and the "shiny" new toys, but don't have the experience to evaluate which of these technologies is right for the organization. Then it becomes *your* job to investigate, and quickly!

Even though your boss has asked you to investigate technology, this could be a great opportunity to create a prioritized "bare bones" artificial intelligence roadmap based on the value to the organization. But how do you go about doing that in just the seven days you've been given?

Wouldn't it be useful if you had a step-by-step guide, a methodology if you will, to guide you through the process? Well, luckily, you've found it here in five steps!

The AI Roadmap

The methodology discussed below, not only works for you as an employee of an organization, it also works in different contexts—you may be a consultant advising your client on AI or a software or services vendor trying to show the value of your AI solution to a customer. How you package and deliver the output of this methodology may differ, but the process will remain the same.

Step 1: Analyze What Matters to the Organization. Grabbing the KPIs

Every organization is slightly different in what drives their existence. Some organizations are driven by the creation of profits to enrich their shareholders, others by the ability to serve citizens of a community, others are dedicated to charitable activities.

Commercial organizations, whether a financial services company, an online store that sells mobile phones, a timber wholesaler, or an airline, are driven by profit.

By way of contrast, government and the not-for-profit sector are, for the most part, not concerned by profit, but by other drivers, such as how effectively they process interactions with citizens or perhaps how many homeless people they can shelter each night. They are interested in the efficiency of delivering their services though, so the cost to deliver each service is always a factor.

All these organizations employ key performance indicators (KPIs) to measure the ongoing performance of the organization. Qlik says that KPI is defined as a "quantifiable measure of performance overtime for a specific objective."

Step one is for you to discover and document the KPIs that are most important to your organization. You may wish to examine the organization's annual report or talk to executives, or you may be able to deduce the KPIs yourself just by observation. In many cases they are obvious, e.g., "revenue per annum."

For instance, if your organization is in telecommunications and sells mobile phones and mobile phone service to the general public, some KPIs could be:

- Cost to create a new online product listing—measured in $
- Average revenue per user per month—measured in $
- Cost to service each query—measured in human time or $
- Average products per user—a numerical value
- Net profit per month—measured in $
- Number of iPhone 16s sold this week—measured as a numerical and/or $

If your organization is a government body concerned with disability services, some KPIs could be:

- Complaints addressed per week—a numerical value
- Number of queries addressed per week—a numerical value
- Employee satisfaction—a numerical value on a scale
- Calls answered within two minutes—a numerical value
- Cost per in-home service interaction—measured in $

When you have collected and understood the most important KPIs of your organization, you should be able to create the necessary output from step 1, which is a list of some KPIs for your organization and how they are measured.

Step 2: Research and Consider AI Use Cases

The second part of building your AI roadmap is to understand what's possible with AI, and how it can be applied to your organization.

Before we get into some depth about artificial intelligence use cases, it is probably useful for us to discuss what different types of AI exist. Contrary to what you may have heard recently, AI has been around for many years! In fact, this author wrote his first paper on AI while at university in 1990 (for which he got an High Distinction, may I add). AI is embedded everywhere, it's in your computer, it's in your phone, it may even be in your cooking appliances. It's not actually that new.

What is new is newer technologies and large-scale use. There are many definitions of AI, but most of the AI used in organizations fall into one or more of these categories:

- *Machine Learning*: AI that "learns" from data using statistical analysis and can be used to predict future outcomes, e.g., predicting sales of bananas based on many different types of data.

- *Natural Language Understanding*: including speech and text recognition, e.g., interpreting language of humans and acting upon it.

- *Conversational AI*: the use of a variety of artificial intelligence techniques (including natural language processing) to understand, process, and give responses to human languages.

- *Generative AI and Large Language Models*: AI that produces various types of synthetic content, such as images, videos, data sets, articles, text, poems, etc. (ChatGPT is a generative AI).

- *Computer Vision*: AI interpreting information from videos or photographs.

Any or all of these types of AI may have a place in your organization. The layer above the technologies is "use cases," which combine one or more of the AI technologies above. Some might have no place in your organization as they do not form part or all of a use case appropriate to your organization.

You may already have some high-level notions about how your organization could use AI. Perhaps "We could automate some of the customer-facing processes. A chatbot to automate the answering of customer queries," "We could run AI over our existing donations data to predict which charity donors are going to donate and when," or "AI could help our legal department in the creation of large legal documents."

The simplest way of investigating how AI could be used in your organization is to investigate what other organizations have done to date with AI. And there are now plenty of examples to be found with some brief research on the internet.

It is possible to find use cases that are specific to your organization's industry, but more common is finding use cases that are generic or pertinent to many different types of organizations. For instance, customer service chat bots on the website will be common to financial services, retail e-commerce, or government departments, in fact, almost any type of organization.

Other AI use cases can be quite narrow and quite specific to one type of organization. We are all familiar with AI-driven product recommendations on Amazon.com, "If you liked this, you'd probably be interested in these products." That type of retail product recommendation is unlikely to be necessary for a government department of births, deaths, and marriages!

To find AI use cases that would be appropriate to your organization, you could try some internet search terms, using the search phrase "AI use cases in [insert your industry here]." For example: "AI use cases in financial services," "AI use cases in the contact centre," "AI to optimize sales," "AI delivery of government services," etc.

The output from step 2 should be a shortlist of candidate artificial intelligence use cases that could be pertinent to your organization.

Step 3: Evaluate the Value of Each of the Use Cases to the Organization

In this step you will evaluate the impact of each shortlisted use case on the KPIs of the organization.

During your research of the use cases, be sure to find numerical values of the use of the AI that has impacted the other organizations' KPIs.

You could find many examples of the impact of AI on KPIs, like these:

- Western Sydney University saved 30 minutes a day per customer service agent by using an AI platform, thus increasing productivity and potentially lowering staffing cost KPIs.
- Minijob-Zentrale saved 75% of the time it takes to write articles using generative AI.

There are plenty of examples like this, with details of the impact on KPIs across all industries including government and not-for-profit. Walmart is a great example. They publicise a great deal about their AI innovations. The Walmart use cases could be used across many industries, not just retail.

With that level of detail, you should be able to make some detailed calculations about how the use case could provide value to your organization. You could provide these projected values as part of your overall report.

If we use the examples above, Western Sydney University saved 30 minutes a day per customer service agent. You can calculate, for this use case, that it would positively impact the productivity of the customer service agent in your organization. But you would need to understand the total number of agents, how long they work, and how much they cost.

Here is another example based on the generative AI example use case above involving Minijob-Zentrale. In this example, we're using some fictional values, targeting a KPI of cost to produce articles:

Articles written per week: 57

Time to write each article: 6 hours

Cost per hour of a writer: $75

Each article costs to write: 6 hours x $75 per hour = $450

Calculated cost per article after 75% saving: $450 * 0.25 = $112.50

Calculated savings per article: $450 - $112.50 = $337.50

Calculated total cost savings per week: 57 * $337.50 = $19,237.50

Calculated total cost savings per year: $1,000,350

This calculation shows a very significant cost saving in the production of articles. You could certainly describe this result as "high impact" as it is making massive savings.

If you haven't been able to find absolute values of the numerical impact of use cases in other organizations, that's okay. You can make some assumptions and use judgement about the level of impact that that use case would have on each chosen KPI. You can specify "high impact," "low impact" or "moderate impact."

You will need to dig deeper though. This high level of savings and productivity gains in the Minijob-Zentrale example, while having a positive financial impact, could result in a negative impact on employees as the result of layoffs or redeployments. AI can also have a positive impact on humans by making jobs more "interesting" by allowing them to concentrate on less mundane and repetitive tasks. The impact on employees can be a complex consideration but needs to be included in your report.

The output from step 3 is a list of your organization's AI use cases, with some annotations about how each use case would impact each KPI. Plus, include some indication of how the use case would impact employees in a positive or negative way.

Step 4: Evaluate the Cost and Effort of Each Use Case

As well as researching appropriate AI use cases, it is important to understand the effort that your organization will need to exert to implement the AI initiatives. A use case could save (let's say) hundreds of thousands of dollars, but it could cost many millions and two years to implement. That would be a really bad thing!

Some use cases are straightforward and low cost, as the tools are already available from the software vendors your organization already uses. Maybe it's just a matter of turning on those AI features. Let's call this type of "everyday AI" "low complexity."

Other use cases could require a much larger effort and cost. Implementing a custom generative AI large language model (LLM) over a corporate document repository would be truly transformational but perhaps an enormous job. Let's call this type of "transformational AI" "high complexity."

If the level of effort is not immediately obvious, you may wish to consult with the right department or staff from your software suppliers. They should be able to help you.

You may need to make a judgement call on where any given use case sits between "low complexity" and "high complexity." You can decide in the next step, when it comes to mapping the results out visually. You could choose to use a numerical range of 1 to 10.

The output from step 4 is a list of your organization's AI use cases, with some annotations against each showing and the effort required to implement each. This will be a range, perhaps from "high complexity" to "low complexity."

Step 5: Represent and Present Your Results

This is the fun part! Now it's time to get ready to present the findings to your manager Graeme. Hopefully, by using this methodology it's taken you less than the seven days specified, and he will be super happy!

In the simplest possible terms, there are two dimensions to the findings you have made. For each of the shortlisted use cases you now have some idea about how much positive impact each would have on KPIs, as well as some idea about the complexity and costs of implementing each use case.

A great way of representing your results is in a quadrant matrix. Executives are usually quite familiar with this form of presentation. A quadrant matrix shows the best results in the top right corner and the worst results in the bottom left. In this case you'll use a quadrant matrix to plot the benefits of the use case to specific KPIs (on the Y axis) over the cost and complexity of each use case (on the X axis). You can produce as many or as few of these as you wish for any number of KPIs. As many as Graeme needs to be happy!

Ideally, from previous steps, you will have been able to get some actual values for the impact of particular KPIs, for instance, implementing "AI-driven product recommendations" on the website would increase your sales revenue by 10% per annum. You could use these actual values on the Y axis if you wished.

As an example, if your organization is in the retail telecommunications industry, one of their very important KPIs is the rate of retention of contracted customers. You could create a quadrant matrix chart plotting the projected impact of each of the AI use cases on the rate of retention of customers KPI.

Example "Impact on Customer Cancelations vs Complexity" Priority Matrix

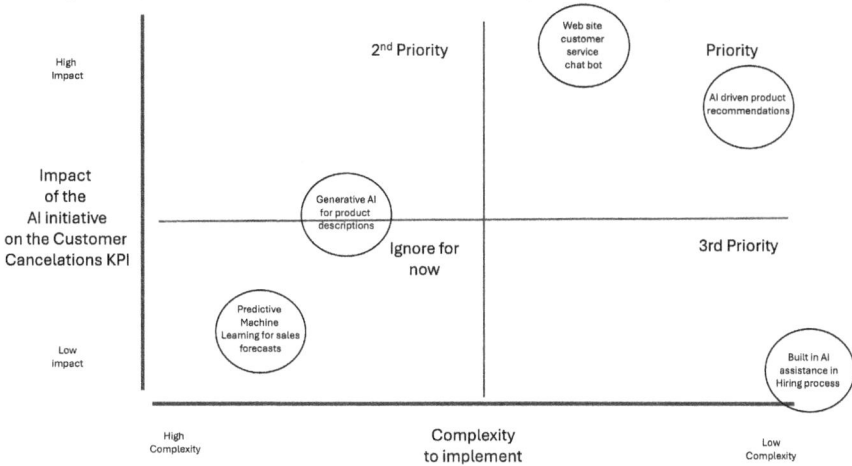

In the example above you can see that two of the use cases have fallen into the "priority" segment of the quadrant, as they are "low complexity" and "high impact" on the retention KPI. This means these use cases will be the obvious higher "bang for buck"—the most impact for the smallest effort. While others fall into the less attractive segments of the matrix and are a lower priority for the organization. The organization should, of course, on the face of it, implement the "priority" use cases first and leave the others until later. Note—when you consider other KPIs, this may change the priority of a given KPI.

If you were to choose a different KPI and plot that against the same use cases on a new quadrant matrix, the results could be quite different. The example below shows the impact of exactly the same use cases but against the revenue KPI.

Example "Impact on Revenue vs Complexity" Priority Matrix

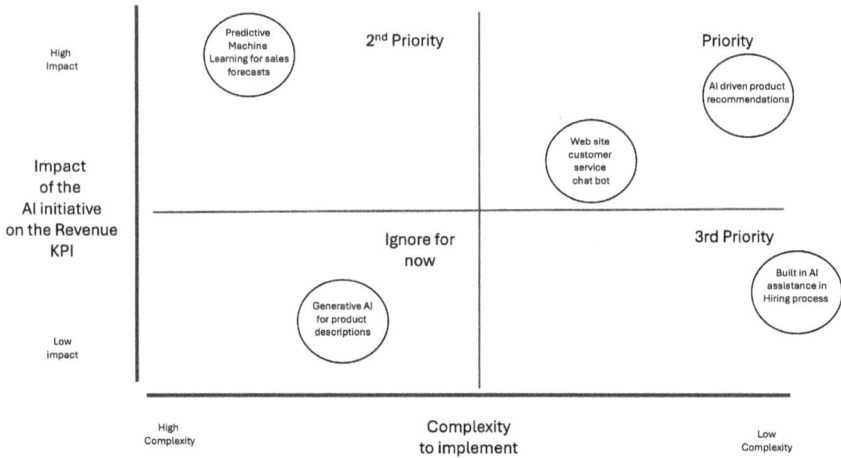

As well as considering the use cases and their impact on KPIs, it's important to consider the impact of the use cases on other stakeholders such as the employees in the organization. If the use of AI was to create a significant loss of jobs in your organization, that's a significant negative impact on the employees and needs to be taken into consideration in your recommendations (although Graeme may see staffing costs as a KPI to be reduced!). The use cases also have a positive impact upon the experience of employees by making their jobs more productive or more fulfilling.

The example below shows how a quadrant matrix can be used to show positive and negative impacts on employees. Again, the results of this analysis could be quite different from the results of the analysis of KPIs but still form important parts of the complete story to be communicated to your manager.

Example Employee "Impact vs Complexity" Priority Quadrant

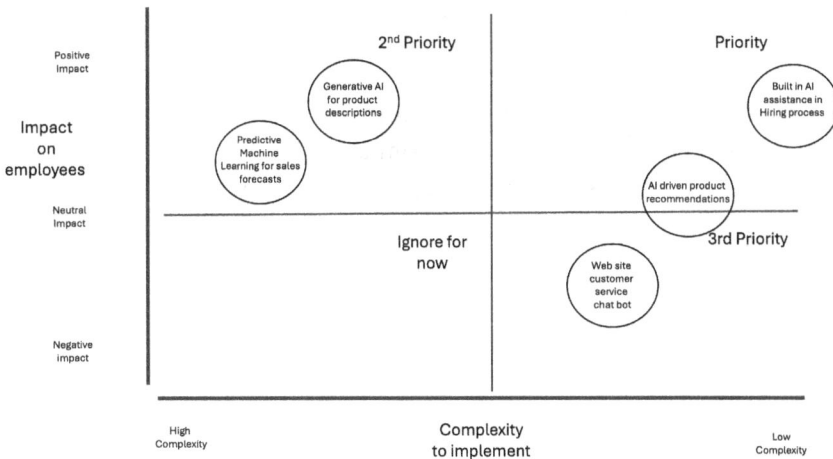

In presenting these results to your manager, you can present multiple quadrants like these showing how artificial intelligence use cases will impact the different KPIs and the impact on other stakeholders such as employees.

Based on the feedback from your manager and the other executives, you can quickly form a list of prioritised use cases to implement in your organization.

The list now forms the start of your organization's artificial intelligence roadmap. At least you can now go forward with confidence, knowing that you've investigated what artificial intelligence to implement, in what order, based on the value to the organization. You have also measured the human impact as part of the consideration.

No doubt Graeme will now ask you to go off and implement it yesterday. Good luck.

About the Author

Carl Jones is a 30-plus-year veteran of the software industry. He has lived and worked in Europe, North America, Asia, and Oceania. He has predominantly worked for large software vendors, which has raised his interest in showing the business value of IT solutions to organizations in all industries. Carl has an MBA from the Australian Institute of Business in Adelaide, South Australia, with a specialisation in Entrepreneurial Management. He lives in Melbourne, Australia with his wife, Debbie, and his Jack Russell terrier, Woody. He and Debbie are adventure travelers. Just this year they explored the fantastic Falkland Islands in the South Atlantic. Carl has visited all seven continents twice.

Email: contact@valuevisionaries.com.au
Website: www.valuevisionaries.com.au
LinkedIn: https://www.linkedin.com/in/carlgjones/

CRAFTING YOUR AI STRATEGY: A HUMAN EXPERIENCE

By Kerry Kurcz
AI & Data Strategy Consultant
San Francisco, California

All models are wrong. Some are useful.
—George Box, 1976

What Is AI?

AI is a tool which leverages data and mathematically powered instructions and parameters for an output. Father of statistics George Box reminds us to use AI for what it is—a tool. While he did not witness generative AI, it is a sobering thought to recall that no model can perfectly represent reality; it is only valuable when the output is insightful for human decision-making.

In Lake Tahoe, California at 12:15 am, I paused during a relay race to look up at the night sky. It was August 2024. A blue moon and billions of stars illuminated the celestial sphere. With a little imagination, one star connects to another and creates a constellation, like Aries. If you know how to find Hamal, the brightest star in Aries, you can find the constellation itself.

The night sky mirrors the mind. The mind mirrors AI. One star connects to another star. One thought connects to another thought. One neuron meets another in a neural network. Thoughts will arrive at an endpoint given boundaries and parameters. The parameters, the algorithm, and the data will guide your AI to where it needs to be.

What Is Machine Learning?

Coined by John McCarthy in 1956, AI is just software—software that can enable a machine to sense, reason, or act like a human. Traditionally, software algorithms were instructive—turn right, turn left, go up, go down, etc.

Machine learning (ML) introduces the opportunity for the *data* to help instruct the algorithm using parameters (source: https://cloud.google.com/learn/artificial-intelligence-vs-machine-learning). One can choose a parameter and then follow down that path, see if it yields good results, and if not, it starts over at a different starting point. Like finding the star Hamal among the billions of others and then seeking Aries, parameters are a guide with a "best guess" starting point. ML shifts from traditional algorithms using experiments which leverage the data and various parameters rather than explicit instruction.

Understanding Your Data

A model depends on the data. During my early studies at Northwestern's AI graduate program, we were taught that data did not necessarily mean insight, nor even information. Certainly not intelligence. Data must be cleaned, wrangled, munged, prepared, normalized, standardized, and understood before it could become anything close to useful for profitable decision-making. This was before the AI boom of 2022, but the concept remains crucial, for three reasons:

1. First, it is essential to understand what data the AI has been exposed to. Has it been ingested in its raw form? Has it been preprocessed, transformed, or cleaned? If it has been altered, what specific changes have been made? Are these preprocessing

steps and transformations documented? What makes the data noisy? What patterns do we see, and can we explain them?

2. Understand the data is both quantitative and qualitative. Where did it come from, is it accurate, and who created it? How many records do we have, how many fields, is it structured or unstructured, and how can we make it more structured?

3. Given we know the data we are working with, we can now begin to apply it to our business problems. What do I think I can solve with this data? What is the data trying to tell me? This is one aspect of what makes AI, AI—letting the data speak for itself. AI will not expand to anywhere beyond the data it has been exposed to. Hallucinations are an inappropriate application of the data it has been ingested with—not a formulation of new thought. That—for now—is reserved for humans, and it acts as an opportunity for you to add value to the business.

Using AI Mindfully

According to the US Energy Information Administration, 2,500 megawatt hours is just about enough to power the average American home for 250 years. It is also the amount of energy required to build a large language model in 2023. Besides the pure energy to keep the lights on and the servers running, machines also require additional resources such as water to keep cool. It is imperative to keep mother nature in mind when inquiring a chatbot to produce work.

As AI can help us with more logical tasks, our human abilities are more important than ever. It is uncertain how the use of AI will affect the next generation's creativity and other skill sets. I am sure my parents were worried about how the internet and spell check would affect mine. Now that we know Millennials can't spell, will Generation Alpha become infamous for their inability to write? Outsourcing spelling is one thing; outsourcing thinking is dangerous, as writing requires thinking. Audio, images, videos, and text, be it AI-generated or not, can be misleading. Challenging yourself also maximizes your chances of thinking for yourself and combating against falling for any of these forms of misinformation, disinformation, or malformation. More and more studies demonstrate how exposure to nature can increase creative and innovative thought, emotional intelligence, and improve mental health (https://www.unr.edu/nevada-today/news/2023/atp-nature-and-the-brain).

The misuse of data and the increasing commonality of data breaches highlight the importance of keeping data secure in creative and innovative ways. Supercomputers which can decode complicated encryption keys have influenced many companies with larger budgets to go beyond the basic security measures. Organizations like the Federal Trade Commission (FTC) also provide guidance on handling data breaches (https://www.ftc.gov/business-guidance/resources/data-breach-response-guide-business).

The Power of Cloud Computing and Multi-Modal Models

Some historically tricky problems, like Fermat's Last Theorem, have been solved in modern times by human thought alone using theoretical mathematics, rather than computational brute force (Simon Singh's *Fermat's Last Theorem*). For other problems, it would take humans multiple generations to solve—where an optimized AI tool can calculate them in seconds. For example, brilliant minds like Isaac Newton considered the n-body problem but were unable to complete it beyond two bodies in the 17th century, due to the complexity and non-linear nature of the equations involved (https://sites.math.washington.edu//~morrow/336_11/papers/loc.pdf). The n-body problem involves predicting the individual motions of a group of celestial objects interacting with each other gravitationally. Specifically, given the initial positions, velocities, and masses of n bodies, the problem is to determine their future positions and velocities over time. Given modern computing power, scientists can now solve the n-body problem for very large numbers of bodies.

AI can now create presentation slide decks, images, video, text, and more. It can write reports, and it can create the below pie chart, representing the market share of the biggest players of cloud computing. Tools like Anthropic's Claude and Google's Gemini can give code to create a pie chart like this. With further prompt engineering, they might even draw it for me. Meanwhile, both models alike failed to let me know that no one uses pie charts anymore.

Cloud Compute Market Share

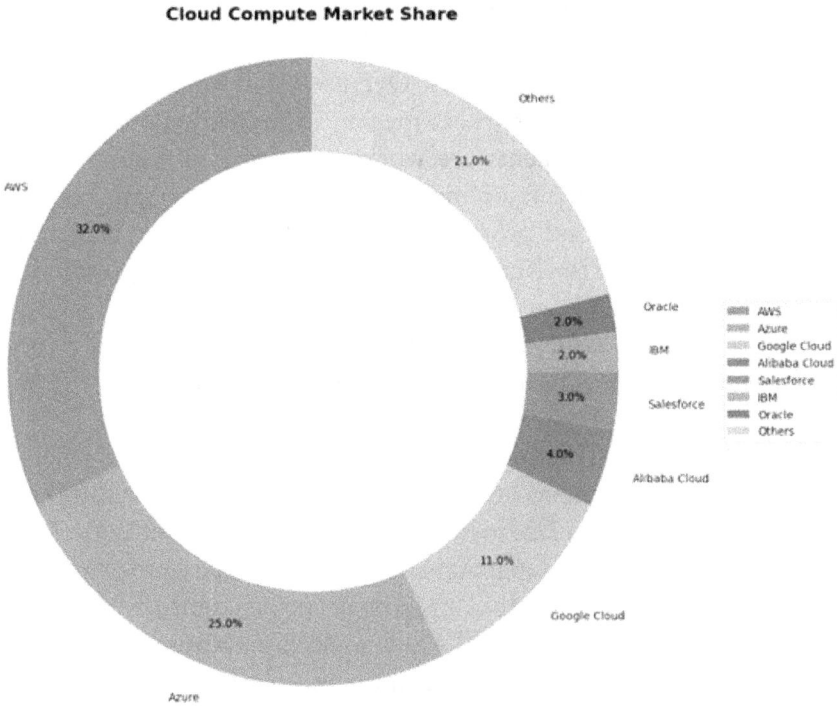

(Statista https://www.statista.com/chart/18819/worldwide-market-share-of-leading-cloud-infrastructure-service-providers/ and Claude3 by Anthropic)

The complicated concept of cloud computing can be simplified to this: rather than buy a computer and be obligated to that purchase until its retirement, you can rent them as you need, paying only for what you use. It's great! But if your data is proprietary, precious, sensitive, and/or confidential, you must understand where those computers are physically located, who owns them, and other data governance, privacy, and security concerns.

Data Privacy, Governance, and Security

The value of data brings significant responsibilities. For example, how can we differentiate between human- and machine-generated content? Understanding the source of information is as imperative as the source of the data that produces it.

In today's digital age, data has become a crucial asset for organizations, driving not only decision-making and innovation, but competitive advantage. OpenAI, the owner of ChatGPT, maintains data partnerships with companies like Time Magazine, The Atlantic, Reddit, News Corp, Vox Media, Financial Times, Stack Overflow, and more (https://openai.com/news/company/). As markets race to capitalize on their data and compute power, governments are working toward AI regulations, balancing the encouragement of innovation with maximizing regulatory action for the safety and protection of all. The United States has published the NIST AI Risk Management Framework (RMF) and industry self-regulation. The Cyberspace Administration of China (CAC) published its Interim Measures for the Management of Generative Artificial Intelligence Services. The European Union has proposed the Artificial Intelligence Act.

The New York Times filed a lawsuit against OpenAI and Microsoft for copyright infringement because it suspects its data was used (https://www.axios.com/2023/12/27/nyt-microsoft-openai-lawsuit-copyright-infringement?utm_medium=partner&utm_source=microsoft-start&utm_content=link&utm_campaign=subs-partner-msoft-login). While it never made it to court, OpenAI's Sky sounding alarmingly like Scarlet Johanssen could have been the reason the product was dropped (https://www.georgetown.edu/news/ask-a-professor-openai-v-scarlett-johansson/). Indeed, generative AI has raised questions like, is my data and my generative AI content eligible for copyright? And is web scraping legal? The answer to these questions is constantly changing and dependent on lawsuits such as Getty Images v. Stability AI, Authors Guild, Inc. v. Google, Inc., LinkedIn v. HiQ, and Andersen et al. vs. Stability AI Ltd. et al. Meanwhile, the verdict of these lawsuits can only apply to so many jurisdictions.

While the United States and the rest of the world provide guidelines to handle generative AI with respect to the law, data privacy is something most, if not all, companies are affected by today. Public figures like Lena Khan have advocated for stronger data protection policies and regulations, reflecting a broader societal shift towards prioritizing user privacy. You might notice there is an overwhelming amount of content to understand how to adhere to data privacy laws, globally. Here are the top 10 data privacy acts to be aware of:

Act	Country	Website
General Data Protection Regulation (GDPR)	European Union	https://gdpr-info.eu/
California Consumer Privacy Act (CCPA)	United States (California)	https://oag.ca.gov/privacy/ccpa
Health Insurance Portability and Accountability Act (HIPAA)	United States	https://www.hhs.gov/hipaa/index.html
Personal Information Protection and Electronic Documents Act (PIPEDA)	Canada	https://www.priv.gc.ca/
Data Protection Act 2018 (DPA)	United Kingdom	https://www.gov.uk/data-protection
Brazilian General Data Protection Law (LGPD)	Brazil	https://lgpd-brazil.info/
Personal Data Protection Act (PDPA)	Singapore	https://sso.agc.gov.sg/Act/PDPA2012
Australia's Privacy Act 1988	Australia	https://www.legislation.gov.au/Series/C2004A03712
Japan's Act on the Protection of Personal Information (APPI)	Japan	https://www.ppc.go.jp/files/pdf/Act_on_the_Protection_of_Personal_Information.pdf
South Africa's Protection of Personal Information Act (POPIA)	South Africa	https://popia.co.za/

Defining what data needs protection is the first step to adhering to data privacy protection. Each of the above acts defines sensitive or personally identifiable information (PII) in various ways. To understand what PII is on a field-level requires understanding the entire dataset.

A field by itself may not be considered sensitive, but when combined with another field, it becomes PII. For example, birthdate alone will not identify an individual. "Birthday" in combination with other fields such as "name" will become more sensitive. Therefore, physically or digitally storing this information separately may help mitigate risk. Having internal policies of classifying and managing PII which take any relevant regulatory act into account is the best way to assure clients that their data is safe and protected.

What Is Natural Language Processing and What Makes a Large Language Model (LLM)?

So far, we've described AI as software to behave like a human, and machine learning (ML) as AI fueled by data. Deep learning is a subset of ML, where sufficiently large datasets can produce an output based on a design inspired by the human brain, such as the neural network (https://aws.amazon.com/what-is/deep-learning/). Natural language processing (NLP) spans across all three.

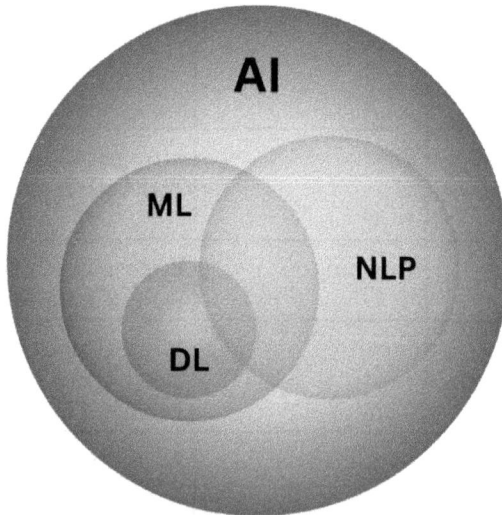

(Detection of Offensive Language in Social Media Posts—Scientific Figure on ResearchGate. Available from: https://www.researchgate.net/figure/Relationship-between-AI-ML-DL-and-NLP-7_fig8_343079524 [accessed 14 Sept 2024])

In the 1960s, Frederick Mosteller from Harvard worked with David L. Wallace from University of Chicago to successfully analyze and classify anonymous pamphlets using word frequencies. Deciphering which author wrote what pamphlet was a problem that had scratched the heads of historians for ages. Thanks to the NLP work of Mosteller and Wallace, the anonymous pamphlets have been classified as either the famous John Madison or Alexander Hamilton with considerable certainty (DOI 10.1080/01621459.1963.10500849).

In addition to word frequencies, NLP may also mean optical character recognition (OCR). Lawyers who used to physically go through papers and papers of documents, redlining and highlighting important phrases, can leverage OCR to automate this process. Any type of NLP still depends on the data. Using the legal industry as an example, lawyers now rely heavily on legal research databases to obtain case law data, statutes, regulations, legal commentary, and more. These digitized tools revolutionized the law industry by enabling those who were willing to upgrade from boxes of papers to digitized, easily searchable texts. The power of automated OCR was only available to those who digitized their library, since the ability to scan documents became more commonplace in the 1990s. Indeed, maintaining at least one digital copy of all documentation no longer appears to be an option for those who wish to stay competitive–especially in law.

Gradually, we have become more sophisticated in our ability to vectorize text, taking the ability to apply AI on proprietary data beyond simple OCR. Vectorizing text for the machine to ingest can be as simple as turning documents into word frequencies or as complicated as leveraging the transformer model (Google, 2017) to create word "embeddings." Embeddings are special in that they consider the context of a word—for example, with the powerful ability to differentiate the word "bank," as in, "I went to the bank for money" versus "I went to the riverbank." Embeddings stored in pricy but efficient vector databases are often used in RAG models, semantic search, or your typical Q&A chatbot. Embeddings are commonly created and updated using state-of-the-art large language models (LLMs).

How to Measure AI Results

As the manual labor to obtain documents is becoming more automated, the need for both innovative and sound human judgment is more imperative than ever. While metrics can suggest various levels of model accuracy, only a human can decide whether a model is useful for their needs.

Moreover, accuracy will not always be the best way to measure success. In classical machine learning problems on imbalanced datasets, recall and precision are better. Focusing on precision implies you want to minimize the false positives. Recall implies you want to minimize the false negatives. F1-score can be used to balance both.

For generative AI output, measuring quantitative performance can be tricky. Testing on a subset of the larger dataset can help to maintain a more feasible testing strategy—and faster deployments. Once validated manually on a smaller dataset, the model can instill confidence in users for a larger dataset. When something goes wrong, it is easier to pinpoint the issue.

Finally, providing transparency on what data is being used, how it may have been modified, and how the models were validated is essential. Without this, why should anyone trust its output?

What AI Does Well

AI is like that super-efficient, slightly too enthusiastic intern who never needs coffee but might accidentally start a paperclip maximization apocalypse if not watched closely. Here's what AI excels at:

- *Pattern Recognition*: AI can sift through oceans of data faster than you can say "data breach." It spots trends and anomalies in datasets so vast, they'd make your spreadsheet crash.

- *Content Generation*: from crafting sonnets that could make Shakespeare roll in his grave (in laughter or horror, we're not sure) to creating images that might just win an art contest (if the judges were robots), AI is your content creation wizard.

- *Automation*: got a task that's as exciting as watching paint dry? AI can do it, and probably better. It's the ultimate in repetitive task automation, freeing you up for ... well, more interesting things like existential crises.

- *Enhancing Human Capabilities*: AI isn't just about replacing; it's about enhancing. Think of it as a sidekick, not a replacement. It can analyze, predict, and suggest, making you look like a genius in meetings.

- *Learning and Adapting*: AI learns from experience, which is more than can be said for some of your exes. It adapts, improves, and sometimes, if you're not careful, it might just start questioning its existence.

- So, what does AI mean to you? Is it a tool, a threat, or perhaps a new form of life that we're just beginning to understand? Think about it, but maybe not too hard; you might end up writing a sci-fi novel instead of finishing your report.

This section was written with assistance from Grok 2 mini (beta –Fun mode).

What AI Still Doesn't Do

AI does not get things 100% right all the time. That is mostly because nothing and nobody does. For deterministic models, the process is rule-based and does not change; this is where AI can excel. For probabilistic models, where the decision is based on a confidence score above some threshold, it may be useful to review the mediocre or low confidence predictions manually and allow the AI model to automate the outcomes with higher confidence. It may also be wise to keep a human in the loop, supervising the AI, until it is ready for complete autonomy.

According to the US Copyright Office, AI does not produce original content. At least, AI-generated content is not eligible for copyright as of the time of this writing (https://www.mofo.com/resources/insights/230301-copyright-office-denies-claim).

AI does not currently seek personal growth. Value systems humans have in place are much different than those of AI. AI also currently does not actively seek knowledge outside the dataset it knows. Although some like Perplexity can search the web when queried for an answer, AI is only exposed to what the humans provide it. Even if the AI can scan the web, there is plenty of knowledge and context that humans have that the AI may not understand, like gravity.

AI alone should not justify investment costs without a business problem to support it and data-driven evidence to support the claim that AI can help solve that business problem. A fast and simple proof-of-concept may be the best way to provide that evidence.

Meanwhile, AI is changing and updating quickly. It will be interesting to see how top companies race for the most state-of-the-art model.

Yet even the most sophisticated AI model is merely a tool to become better at our jobs and, as humans, add value to the world and grow.

About the Author

Kerry Kurcz is an AI & Data Strategy consultant in San Francisco, California. With an architect's mindset and extensive experience in data, AI, & product with companies like J.P. Morgan and Obviously AI, she holds a master's in data science from Northwestern University, specializing in AI & NLP. Passionate about data quality, strategically leveraging technology for innovation, and bringing siloed teams together, Kerry has led dozens of transformative AI projects and developed impactful solutions that enhance the human experience. Recognized with the Rising Star award from J.P. Morgan, she has shared her expertise as a speaker at Northwestern and Norwich University and her published works contribute to advancing AI research. Outside of work, she enjoys cycling, swimming, running, hiking, music, baking, reading, and travel, which inspire her creative thinking in AI.

Connect with Kerry:
Email: kerry@kb3k.com
Website: kb3k.com
X: https://x.com/kb3labs
YouTube: https://www.youtube.com/@kb3labs
LinkedIn: https://www.linkedin.com/in/kb3k

THE FUTURE OF DEFENSE INNOVATION WITH AI AND BLOCKCHAIN

By Rudy Martinez, MS
Digital Humanities Explorer, Doctoral Candidate
Dallas, Texas

> *The future influences the present just as much as the past.*
> —Friedrich Nietzsche

Picture a near future, where cutting-edge AI and unbreakable blockchain technology combine to create a fortress of security, a digital sentinel guarding our most critical assets with unparalleled intelligence and integrity. In this world, every operation is flawless, every decision is precise, and every system is beyond compromise. Welcome to the future of defense innovation, where the synergy of AI and blockchain redefines what is possible.

September 17, 2043

The hum of the drone's rotors cut through the stillness at dusk. Below, the urban landscape was a mixture of chaos and quiet, streets filled with the remnants of a conflict that had lasted too long. The drone, a sleek and advanced piece of technology, moved with surgical precision, its sensors scanning for signs of life amidst the rubble and debris.

Equipped with state-of-the-art cameras, infrared sensors, and environmental scanners, the drone was a marvel of modern engineering. It hovered above a particularly devastated area, its AI-driven algorithms processing the incoming data stream in real time. Infrared sensors detected several heat signatures trapped inside a partially collapsed building. The drone's onboard AI, cross-referencing the data with its extensive database, quickly identified them as civilians in need of rescue.

"Operation Safe Passage, initiate," commanded the synthesized voice of the AI overseeing the mission. The drone's systems sprang into action, each move calculated with precision. Micro-drones were deployed, dispersing into the air like a flock of birds, each equipped with high-resolution cameras and sensors. They formed a mesh network, enhancing the primary drone's view and feeding data back to the command center.

The drone's loudspeaker crackled to life. "Attention, civilians. This is an autonomous rescue operation. Follow the illuminated path to safety." On the ground, a holographic path appeared, projected by the micro-drones, guiding the trapped individuals towards the nearest safe exit. The drone's sensors identified armed insurgents blocking the primary escape route. Utilizing non-lethal measures, the drone deployed incapacitating gas canisters, neutralizing the threat without causing permanent harm.

The civilians, now safe and guided by the holographic path, moved cautiously through the war-torn streets. Their faces, illuminated by the glowing pathway, showed a mixture of fear and hope. They followed the lights with the trust that the unseen guardian above would lead them to safety.

So What Exactly Is AI?

Artificial intelligence (AI) is like the magic wand in the realm of technology, transforming ordinary machines into thinking, learning entities. Imagine a robot with the curiosity of a detective, the memory of an elephant, and the problem-solving skills of a seasoned chess grandmaster. AI is the secret sauce that allows machines to learn from experience, much like a superhero gaining new powers after each adventure.

For instance, just as a child learns to recognize a cat by seeing different cats, AI can identify objects by analyzing countless images. It's like giving a computer the eyes of an artist, the ears of a musician, and the brain power of a seasoned explorer, all working together to perform tasks that once only humans could do. From personal assistants like Siri and Alexa who chat with you and play your favorite songs, to self-driving cars that navigate bustling streets like skilled pilots, AI is the enchanting force turning science fiction into everyday reality.

To power the magic of artificial intelligence, robust hardware is essential, especially high-performance computing (HPC) systems. These powerful machines consist of numerous processors and vast amounts of memory, working together to handle complex calculations and massive datasets at lightning speed. Think of HPC as the muscle behind AI, providing the computational strength needed to train intricate models and run sophisticated algorithms.

To see just how much more computing power an HPC has over a single high-end laptop the following table compares each:

Feature	High-End Laptop	High-Performance Computing (HPC) System
Processing Power	~10 teraflops (combined CPU & GPU)	~1 petaflop (1,000,000 gigaflops)
CPU	Intel i9 or AMD Ryzen 9	Multiple high-performance processors
GPU	NVIDIA RTX 3080 or equivalent	Multiple GPUs like NVIDIA A100
RAM	16–64 GB	Several terabytes (TB)

Storage	1–2 TB SSD	Petabytes of high-speed storage
Cooling System	Standard laptop cooling system	Advanced liquid cooling
Interconnects	Standard Ethernet/Wi-Fi	High-speed interconnects (InfiniBand)
Power Consumption	100–300 watts	Several megawatts
Parallel Processing	Limited (few cores)	Thousands of cores optimized for parallel processing
Physical Space	Portable, compact	Large data center required
Cost	$2,000–$5,000	Millions of dollars

By a slight calculation, that would mean it takes approximately 100,000 laptops to match that of a single HPC system.

AI, in its early stages, is much like a human baby. At first, it learns basic tasks through repetitive training, much like how a baby learns to crawl before walking. This initial phase involves feeding the AI system with data, teaching it to recognize patterns and make simple decisions. However, unlike humans, AI doesn't take years to mature. It rapidly transforms, akin to Neo from *The Matrix*, who uploads the ability to learn kung fu instantly. With the right data and computational power, AI can quickly evolve from basic learning to performing extraordinary feats, such as predicting future events, understanding natural language, and even creating art. This rapid advancement is made possible by the combination of HPC hardware and sophisticated algorithms, propelling AI from its infancy to superhuman capabilities in a remarkably short time.

Consider DARPA's ALIAS program, which develops autonomous aircraft systems to reduce pilot workload and enhance mission capabilities, or Project Maven, which uses AI to analyze drone footage and identify potential threats. These case studies exemplify how AI is revolutionizing warfare by acting as the critical brain power behind every strategic decision. But beyond these practical applications, AI in defense offers a unique perspective: it transforms data into actionable intelligence, making operations not just smarter but safer and more efficient.

In a fortified command center not far from the city, a team of engineers and military strategists monitored the drone's every move on a wall of high-resolution screens. The system was decentralized and autonomous, with data flowing securely through an unbreakable blockchain. This digital fortress ensured that every piece of information was verified, encrypted, and stored immutably.

The blockchain acted as a ledger, recording every action taken by the drone and its micro-drones. This technology, once primarily associated with cryptocurrencies, had evolved into a cornerstone of security and transparency. Each byte of data, from the drone's deployment to the rescue operation's minute details, was encoded into the blockchain, creating an unalterable chain of trust.

In this world, the blockchain was more than just a record; it was a dynamic network that facilitated real-time decision-making. Every sensor reading, every micro-drone deployment, and every civilian's movement was captured and analyzed. The blockchain verified and validated each action, ensuring that the data was accurate and untampered. This process allowed the drones to operate with a level of precision and reliability that human oversight alone could never achieve.

As the civilians made their way to the extraction point, autonomous vehicles arrived, their routes optimized in real time by the same AI-blockchain network guiding the drones. These vehicles, designed for rapid deployment in hostile environments, whisked the civilians away to a secure location.

So What Exactly Is a Blockchain?

Blockchain technology is like a digital ledger, a modern-day version of an ancient scroll, meticulously recording every transaction in a secure, decentralized, and transparent manner. Imagine a network of knights guarding a treasure chest, where each knight has a copy of the chest's contents, ensuring that no single knight can alter the contents without the others noticing. Blockchain functions similarly by distributing a record of transactions across many computers, making it nearly impossible for any single entity to tamper with the data.

Just as a library catalog keeps track of books borrowed and returned, databases are used to manage and record various types of data. However, while traditional databases are effective, blockchain technology offers a more secure and tamper-proof alternative. A blockchain maintains a permanent, unchangeable record of all transactions. Each transaction is a "block," and as these blocks link together in chronological order, they form a "chain," creating a trustworthy and traceable history of data. This feature makes blockchain invaluable for applications requiring high security and transparency, such as financial transactions, supply chain management, and digital identities.

Here's a comparative look at how blockchain technology enhances the capabilities of traditional databases:

Feature	Traditional Database	Blockchain
Data Structure	Centralized tables that can be modified or deleted	Linked blocks that are immutable once recorded
Security	Relies on access control and security protocols	Enhanced by cryptographic algorithms and consensus
Decentralization	Typically centralized, controlled by one entity	Decentralized, distributed across multiple nodes
Data Integrity	Vulnerable to manipulation by authorized users	High integrity due to immutability and consensus
Transparency	Limited to authorized users	Publicly verifiable records
Auditability	Requires complex logging mechanisms	Built-in, transparent audit trail
Scalability	Scales easily with additional hardware	Requires significant resources to scale
Transaction Speed	Fast, optimized for quick read/write operations	Slower due to consensus protocols
Cost	Lower, based on infrastructure and maintenance	Higher, due to energy consumption and hardware

In 2016, Wells Fargo employees created millions of unauthorized bank and credit card accounts to meet aggressive sales targets. If Wells

Fargo had employed blockchain technology, the fraud could have been prevented or detected much earlier due to several key features of blockchain:

1. *Immutable Records*: once a transaction is recorded in a blockchain, it cannot be altered or deleted. Unauthorized accounts would have been permanently logged, making it easy to trace and identify fraudulent activities.

2. *Decentralized Ledger*: blockchain's decentralized nature means that multiple copies of the transaction ledger are maintained across numerous nodes. Any discrepancy or unauthorized change in one ledger would be immediately apparent, ensuring quick detection and rectification of fraudulent activities.

3. *Transparency and Auditability*: blockchain provides a transparent and publicly verifiable record of all transactions. Regulators, auditors, and even customers could have independently verified the legitimacy of account activities, making it extremely difficult to carry out widespread fraud without detection.

By leveraging blockchain technology, Wells Fargo could have ensured a higher level of security and transparency, making it nearly impossible for employees to create fake accounts without detection. This example highlights the robustness and integrity that blockchain can bring to financial systems, preventing fraud and enhancing trust.

From securing military communications to tracking the provenance and integrity of supplies, blockchain's applications are vast. It ensures that intelligence shared among allies remains secure and verifiable. Projects like Taekion's blockchain-based military data protection and VIA's application for the US Department of Defense highlight the transformative power of blockchain in safeguarding critical data. These real-world implementations demonstrate how blockchain can enhance data security, streamline operations, and provide real-time insights for strategic decision-making.

But the true marvel lay beneath the surface, in the hidden mechanisms driving the mission's success. This was not merely a showcase of drones and blockchain; it was the pinnacle of a decentralized command center powered by AI.

In the last two decades, a series of breakthroughs had redefined the landscape of defense and security. Advances in machine learning allowed AI to process vast amounts of data and make split-second decisions with greater accuracy than any human. Simultaneously, the development of blockchain technology provided an unbreakable foundation for data integrity and transparency.

Combining these technologies, researchers created a decentralized network capable of managing and executing complex missions autonomously. The AI, embedded within this network, learned from each operation, continuously improving its strategies and efficiency. This system operated independently yet harmoniously, ensuring that every action was precise, every decision was flawless, and every operation was beyond compromise.

As the drone returned to its base, the command center's AI reviewed the operation's data, making subtle adjustments and optimizations for future missions. The engineers watched in awe as the AI autonomously updated its algorithms, refining the tactics and improving the coordination between drones and vehicles.

The civilians, now safe, had no idea that their rescue was orchestrated not by human hands but by an intricate dance of AI and blockchain technology. This was the world 20 years from now, where the synergy of these advancements redefined what was possible. The digital sentinel stood vigilant, protecting humanity's most critical assets with unparalleled intelligence and integrity, in a future that was already unfolding before their eyes.

Defense Innovation Using AI and Blockchain

Combining AI and blockchain can revolutionize defense by enhancing data security, streamlining operations, and providing real-time strategic insights. However, this integration comes with challenges and ethical considerations. Implementing these technologies raises questions about accountability and the ethical use of autonomous systems. For instance, who is responsible when an AI makes a life-or-death decision? Balancing innovation with ethical responsibility is crucial. Transparent protocols and oversight mechanisms must be established to ensure these technologies are used responsibly.

The potential impact is profound. AI and blockchain together can create systems that are not only efficient but also resilient against cyber threats. They offer the possibility of decentralized decision-making, where data integrity and operational security are paramount. But as we adopt these technologies, we must also navigate the ethical landscape they present, ensuring that their use aligns with our values and principles.

Standing at the precipice of this new frontier, we face a future where the lines between man and machine blur, and our very security depends on the flawless integration of AI and blockchain. This is not just an evolution but a revolution in defense technology—one that promises unprecedented protection and daunting challenges. Will we harness this power responsibly, or will we falter under its immense potential? As we venture into this brave new world, we must navigate with wisdom, innovation, and unwavering vigilance—critical characteristics of the AI Mindset. The future is both thrilling and terrifying, and it starts now.

About the Author

Rudy Martinez is a seasoned technology professional with over 30 years of experience, currently serving as Head of IT at Shield AI. His expertise spans cybersecurity, blockchain, cryptographic software, enterprise architecture, and complex infrastructure. Previously, at Visa Inc., Rudy played a key role in launching VisaUniversity.com, introducing the first-ever Payments Certification. His career began in the military, where he excelled in cryptography and satellite communications, earning multiple Navy Achievement Medals and the Enlisted Surface Warfare Specialist designation. Rudy's leadership and strategic planning skills have been demonstrated through numerous deployments and high-level roles. He holds a bachelor's degree in information systems and a master's in cybersecurity, and is currently pursuing a Doctorate of Information Technology. At Shield AI, Rudy drives innovation by enhancing cybersecurity, deploying generative AI and high-performance computing stacks, and supporting the company's AI pilot mission.

Connect with Rudy:
Email: rudy@rudymartinez.ai
LinkedIn: https://www.linkedin.com/in/rudymartinezai/

ADOPTING AN ADAPTIVE MINDSET TO NURTURE BUSINESS OPTIMIZATION AND INNOVATION

Dr. Michael T. McClanahan
Graduate Professor, Enterprise AI Sales
Douglas, Arizona

> *I think what makes AI different from other technologies is that it's going to bring humans and machines closer together. AI is sometimes incorrectly framed as machines replacing humans. It's not about machines replacing humans, but machines augmenting humans.*
> —Robin Bordoliopens

Business leaders are on the verge of a game-changing paradigm shift that will shape how competitive advantage is defined and the need for the workplace to embrace a technologically disruptive inclusive operations model. The rise of artificial intelligence (AI) is forcing management to rethink how they plan for future business growth, staff their organization with knowledge-based workers, and embrace the significance of formulating a new business model based on data-driven insights. The fundamental challenge facing today's business leaders is whether they can nurture and sustain ongoing business optimization and innovation in a workplace saturated with disruptive automation.

Management must overcome technical adversity by aligning a mindset that supports an environment where internal and external technology-authored trends mold decisions and infuse industry best practices. Leaders must also aim to harmonize employees and technology in a dynamic workplace that reinforces continuous learning and encourages organizational growth. Enterprise-wide optimization and an innovative mindset are shaped through habitually exuding confidence in sound business decision-making. To achieve this task, the leader must incorporate seven daily habits that balance harmonizing technological innovation and deriving stakeholder value.

Seven E-Habits to Overcome Technology Disruption in the Workplace

AI's pervasive force across the work setting is forcing an unprecedented shift that is upending conventional wisdom on optimizing. This postmodern change is philosophically challenging the status quo of adapting new organizational processes with operations strategies that demand fluid and inclusive information to adjust and shift seamlessly with minimal intervention. The approach demands collaboration from all stakeholders in the operations, not just a new set of directions for employees to follow.

Disruptive technology derived from AI is the enabler to feed information, coupled with knowledge workers who can apply the information seamlessly to adapt to ongoing adjustments in the workplace. According to a recent McKinsey study, apathetic businesses that refuse to embrace disruptive technology as an integral part of their business strategy will lose touch with defining customer experience, lose talent to their competitors, and increase additional risk to the company. Organizational leaders must heed the warning, embrace the uncertainty of technological disruption, and adopt seven critical habits that facilitate an AI mindset that adapts to a postmodern change management framework.

1. Establish a Strategic Vision with Transparent Alignment

AI must be front and center in the organization's overall strategy and how it weaves into its vision and mission. Management must articulate how AI is defined in the organization and how it is integral to a company's objectives and future initiatives. Initially, the message needs to be described as a journey to embracing AI, how each employee is a part of it, and how they influence the implementation. Over time, the message should shift to expanding capability and maturity and the role AI integrates into the workforce.

Employees at all levels need to feel they are contributing to the new shift, and human resources should also emphasize in their recruiting process the role AI and employees have in the company's success in attracting top talent. The approach and implementation plan must be supported throughout the organization through collaboration. Access to a feedback loop to adjust and react to overlooked factors or specific needs from everyone in the organization is essential. Where plausible, crowdsourcing can influence the milestones and priorities into must-have, should-have, could-have, and won't-have (MoSCoW) to encourage active employee participation. Ultimately, the fundamental goals of establishing an enterprise AI Mindset aligned with the company's strategic vision are to promote participation, remain transparent throughout the journey, prioritize the deliverables with the most impact, and execute a clear integration roadmap that adapts to change.

2. Empower Thought Leadership and Inclusive Collaboration

One of the biggest fears of AI in the workplace is the negative impact it will have on the role of management and the workforce in the future. However, the fundamental truth is that implementing AI in the work setting will mandate and encourage future collaboration among employees across the enterprise. Disruptive technology thought leadership will become decentralized, and there will be a need for engaging buy-in and cooperation across all levels of the organization. Change and ideas will emerge organically and not be solely dependent from the top down.

Power and decision-making will become diffused, and data will often challenge authority. Different perspectives are essential for growth and will co-exist. Inclusivity and different viewpoints will arise and form more of an egalitarian organizational culture. The company's culture will create a more engaging environment built around cultural diversity, where data will enable the organization to co-create and strengthen its outcomes.

Empathy, critical thinking, and adaptability will drive how the company demonstrates value to its consumers. The organization's leadership mindset will shift and become content with empowering employees to do more, providing employees with a sense of feeling more valued, responsible, and accountable for their decisions. Ultimately, the shift is more externally focused and enables a more customer-centric commitment to meet unyielding customer demand through a dynamic competitive advantage built upon inclusive collaboration.

3. Embrace Continuous Learning

Today's technological disruption has moved beyond Moore's Law in solely measuring raw computer power. Computational abilities have been replaced with machine learning accelerators, including graphic processing units (GPUs) and tensor processing units (TPUs). Similarly, employees are either embracing or told to espouse ongoing continuous learning in the workplace at a frenetic pace to keep up with disruptive automation and the adoption of postmodern change. Personal and professional growth now depends on staying on the tip of the spear, being relevant in the industry, and creating opportunities to grow and expand one's expertise.

Continuous learning has many advantages for the workforce, including remaining competitive and the need to innovate continuously. Lifelong learners directly contribute and are equipped with timely information and ideas that can fuel the company and support AI systems in developing and nurturing new ideas and solutions. Continuous learning also helps improve overall problem-solving and action-based decision-making by creatively leveraging disruptive technology.

Companies can also attract thought leaders seeking employment in organizations that like to remain relevant and prioritize knowledge. Making disruptive technology like AI a priority in a company can also retain the best and brightest and reduce turnover rates due to the value placed on ongoing innovative thought leadership. Ongoing investment in employee literacy also serves as an agent to overcome disruption through elevating personal growth and arming everyone with the information they need to remain relevant and more resilient against rapid changes. Fundamentally, active learning in the modern workplace encourages the organization to remain fiercely competitive as a team, adapt to ongoing change, and empower inclusive collaboration.

4. Enable Data-Driven Decision-Making

Data-driven decision-making (DDDM) has vastly improved the capacity to improve accuracy and objectivity through analyzing data holistically across the business enterprise. Business forecasting has evolved over the past 25 years from time-consuming manual data gathering and water-fall-based implementation plans to leveraging deep learning to make subtle real-time adjustments based on prevailing conditions. Enhanced precision and risk mitigation-based algorithms allow management at all levels to make informed decisions through real-time information to derive insights that were never possible before the introduction of AI. Pulling disparate data and identifying potential patterns and insights enables strategic leaders to empower critical thinking and unique insights that enhance the business internally and externally.

DDDM also helps to eliminate human bias by weighing facts and removing the reliance on intuition and emotional decision-making. The approach also enables the business to identify and automate routine and repetitive tasks to shift focus to more strategic activities where the employee can add direct value to the company. DDDM can also scale execution for industries like retail to react and respond to changing market conditions, operational setbacks, and customer behavior through automation to keep up with the fast-paced business environment required to thrive in today's marketplace.

Leveraging predictive analytics helps anticipate market trends and customer needs. Internally, it can provide critical insights into optimizing and allocating resources and proactive decisions on operational efficiencies. As a result, the company can leverage these insights and adopt a data-first strategy to incorporate informed decision-making in its operations and drive a forward-thinking business execution strategy.

5. Enhance Presence and Scale-to-Market Conditions

AI has revolutionized an organization's ability to shape and create a go-to-market strategy effectively, and it is imperative for leaders to re-think how it used to be done and embrace how it can now be done. Historical data, lengthy market research studies, and manual analysis using surveys and focus groups limited the researcher's ability to generalize and scope the effort. Precision was lost over time, and the process itself was time-consuming.

Machine learning, through ingesting real-time data and advanced analytics via various sources such as social media, online interactions, and customer transactions, has upended traditional methods. The result is more precise targeting, adapting to market shifts, and driving a personalized offering at scale. Hyper-personalization, chatbots, and virtual assistants not only improve the customer experience and engagement with the consumer but also enable and provide interactions that are both relevant and timely.

Management's challenge in applying the necessary principles to create a successful go-to-market strategy is balancing an approach that can obtain and gather information insightfully without compromising data privacy. All approaches need to be transparent and aligned with the company's values while respecting and acknowledging the diversity and universality of the consumer. An effective AI-driven strategy focuses on delivering scalable market insights and consistently maintains strong ethical practices as an integral part of the approach.

6. Encourage Resilience and Perseverance

In order to have an AI Mindset, business leaders must overcome the traditional top-down strategic thinking model and implementation strategies. To thrive in today's frenetic pace, managers must embrace unique and diverse insights from the data and the team. One of the most effective ways to achieve this is through cultivating a culture of experimentation. Failing fast and rapid iteration encourages systematic improvement and makes the organization resilient through unforeseen obstacles and uncertainties.

Strategic agility across the workforce embraces flexibility and enables the company to pivot on real-time data to counteract unforeseen market shifts. Data is the organization's lifeblood and must be protected through a well-defined data governance framework to ensure full compliance and prevent security breaches. Cross-functional engagements across the enterprise mitigate against silo-thinking and encourage synergy across the company. The goal is to recognize that the company's power resides in the skills and talents of the workforce, which is rallying behind a unified message and an innovative framework that adapts to reliable information continuously.

7. Ethicize the Approach in Word and Action

The notion of AI running rampant with limited controls and leaving into question whether the approach recognizes the value of humanity is a valid concern. The approach must be grounded in management-enforced ethical guidelines and governance built around the company's values. Roles such as ethical officers and committees must also be shaped to oversee the AI strategy and implementation and serve as an empowered voice of the stakeholders. Clear ethical standards must be published and governed to align with the company's values and adhere to country-based legal policies and human resource obligations.

Transparency is crucial, and it is incumbent that all AI systems regarding data sources and decision-making processes be explainable to every stakeholder. The goal is to develop trust among employees, company stakeholders, and, most importantly, the customer to appreciate and understand the basis of how the algorithms operate and make decisions. Educating and offering information on the operation of the tools can also help the overall implementation through clear lines of communication. The overall operational framework governing disruptive technology in the company is meant to drive clarity and balance the security of the organization's intellectual property.

Diversity matters. Assembling a diverse set of resources to make up the teams enables different points of view to be shared and considered. The team should have equal representation in gender, age, ethnicity, and background to ensure inclusive feedback loops are firmly in place. Making ethical standards the most essential habit in the company's AI strategy will enable the entire process to continuously improve, place a high value on sustainability, and encourage committed participation in change from each stakeholder.

Making It a Reality

AI in the workplace forces a significant paradigm shift in how business leaders optimize operations and drive innovation to establish organizational growth. Shaping an organizational strategy built on AI as the centerpiece is no longer reserved for management alone; it is an opportunity for an inclusive cross-functional collaboration of resources committed to the company's values and future growth. Embracing uncertainty yet armed with ethical data, future business leaders will be recognized for their creative contributions by adopting seven habits designed to encourage measurable outcomes and sustain value to the bottom line.

The company must shift from a handful of decision makers, commit to and encourage a decentralized strategic decision model, accept data-driven outcomes, and embrace the critical role of cross-functional teams interacting and working together. Although resistance to disruptive technology may initially arise due to the fear of job displacement, management must instill confidence in the reliability of AI's abilities and the ongoing demand for realignment and postmodern change throughout the organization. Leaders must also recognize that strict adherence to an ethical foundation is the cornerstone to maintaining organizational cohesion and inclusive thought leadership. Optimizing the business with disruptive technological innovation relies on a workforce that flourishes and grows responsibly through embracing an AI strategy that enables everyone to continuously thrive in the disruptive work setting.

About the Author

Dr. Michael T. McClanahan is a highly decorated military veteran who served in the US Army for 11-plus years. He has been a business leader in the technology private sector for over 27 years, specializing in shaping large, technologically disruptive business deals globally for Fortune 500 companies. Dr. McClanahan formally teaches MBA students about AI practices in business and shaping innovative enterprise business strategies. He also helps others to bring their business dreams to life through personal consultation. Dr. McClanahan holds a Doctor of Business Administration (DBA), a Master of Business Administration with a concentration in Global Management (MBA), and a summa cum laude with a bachelor's degree in psychology. Dr. McClanahan is happily married and lives with his wife and two Yorkies in Douglas, Arizona. He is also the father of five wonderful grown children across the United States.

Email: m_mcclanahan@att.net
Website: www.pcbdreamer.com
LinkedIn: https://www.linkedin.com/in/mmcclanahan/

CHAPTER 19

AI AND CYBERSECURITY: NAVIGATING THE FUTURE

By Toby Miller
Cybersecurity Director with AI Experience
Chicago, Illinois

AI is likely to be either the best or worst thing to happen to humanity.
—Stephen Hawking

Overview of AI's Growing Influence

It's 2024 and I'm sure we all have heard of this little phenomenon called artificial intelligence (AI), right? AI has been making a splash in the media since OpenAI opened up ChatGPT to the world almost two years ago. Since then, AI has been transforming industries such as healthcare, banking, and retail, just to name a few. And don't think the impact of AI is limited to just those industries, AI is being integrated into cybersecurity, autonomous vehicles, and even our everyday devices (smartphones, smart home devices).

With this cool technology making such an impact in our daily lives, who is ensuring that it's not going to harm us, that the data being used to train these AI models is not poisoned or tampered with? I can tell you right now it's not Superman although if he is reading this book, please reach out to me. So, who's responsible? It will be left to the cybersecurity professionals.

The Role of AI in Modern Security

AI has been around for a while. Back in 2016, I deployed a tool that used AI to detect possible attacks on a network. The interesting part was that we had to let it sit on the network and "learn" what normal activity looked like. But AI has come a long way since then. Today's AI has massively improved detection capabilities in the cybersecurity world. Many of the tools used in modern computing include either AI or some form of machine learning (ML)—and we need them because the "actors" are using them too.

What do I mean by "actors"? In cybersecurity, "actors" refers to threat actors—individuals or groups who seek to cause harm to systems. Many of these threat actors are well-funded (often nation-sponsored) and are using the same technologies that cybersecurity professionals use to find vulnerabilities in a company, organization, or nation's defenses. Here's the key difference: a threat actor only needs to be right once, but a cybersecurity professional has to be right 100% of the time.

To emphasize how important AI is to the future of cybersecurity and vice versa, according to IBM, organizations lacking AI security measures face an average data breach cost of $5.36 million USD—18.6% higher than the overall average. In contrast, those with even limited AI security experience a significantly lower average breach cost of $4.04 USD million, which is $400,000 USD less than the global average, and 28.1% lower than organizations with no AI security at all (IBM, n.d.). That alone is scary.

So how exactly does AI improve cybersecurity? Let's take a closer look at some key areas where AI makes a difference:

- *Phishing Detection*: ever checked your email spam folder? Those dangerous-looking emails you find there are often caught by filters using AI or machine learning (ML). ML is especially effective here because it excels at predicting or classifying, making it the ideal tool for identifying phishing attempts.

- *Faster Response*: one of the key metrics in cybersecurity is response time—how fast you can act during an incident. AI helps process and analyze large amounts of data quickly, allowing organizations to respond in real time.

- *Understanding and Prioritizing Emerging Threats*: AI's ability to learn means it can help cybersecurity teams stay updated on the latest threats and prioritize the most urgent ones.

- *Operational Efficiency*: AI helps businesses become more efficient by automating smaller, repetitive tasks, which frees up human experts to focus on strategic decisions. This also reduces costs while increasing productivity.

- *Improved Detection*: AI can process large data sets at incredible speeds, enabling faster threat detection and quicker response times.

- *Regulatory Compliance*: with AI, staying on top of data protection and compliance monitoring becomes more manageable, helping organizations meet regulatory requirements with ease.

- *Coding*: surprised to see this on the list? Thanks to generative AI and models like Codex, even non-developers can use AI to write code. I've personally used AI to develop programs that improve both security and business operations.

The Risks of AI

Over the past couple of pages, we have discussed why AI in security is important and we gave some examples of what AI could do to improve cybersecurity. Now we are going to shift gears and discuss some of the risks associated with AI. While there are many risks associated with AI, we are only going to cover some key risks:

- *Bias and Discrimination*: bias in generative AI can be inherited through training data, algorithm bias, and cognitive bias. For this discussion we will focus on training data. As the old saying goes, "Garbage in, garbage out". If the AI model is trained using data that has biases and is discriminatory in nature, then the AI model will be biased and discriminatory.

- *Privacy Concerns*: privacy is a big issue within AI. As a matter of fact, I could probably write a book on privacy issues alone (lightbulb going off). AI processes large data sets that could include all kinds of information such as data from web history, surveillance cameras, and a host of other data. Facial recognition is also a big privacy issue within AI. It's so big that Microsoft recently stated that it has banned police departments from using their facial recognition service through the Azure OpenAI service.

- *Adversarial Attacks*: adversarial attacks are a form of cyber-attack that "fools" the machine learning algorithm to make an incorrect decision. For example, in an adversarial attack it would manipulate the ML/AI algorithm in an autonomous vehicle to not properly identify a stop sign; therefore, the car would run through the stop sign and cause an accident.

- *Data Poisoning*: this cyber-attack is where an actor would manipulate the training data. When the training data is changed or manipulated, it can cause all kinds of problems. If you're a healthcare company and your model has been changed with you knowing about it, it could cause the AI to provide a false diagnosis.

- *AI in Cyber-Attacks*: AI-driven cyberattacks use AI as a tool to enhance or carry out the attack. One example that many people, including my mom (Hi Mom), have probably heard of is deep fakes—the technology that replicates someone's appearance and voice so convincingly that it's hard to tell the difference from the real thing. It's a frightening development. Deep fakes have also been used in business scams, where attackers impersonate CEOs, CFOs, or COOs, sending emails or making calls that instruct employees to transfer money to fraudulent bank accounts.

Future Trends and Directions

As AI continues to grow at such a rapid pace, its role in cybersecurity will expand as well. This will provide us with new challenges and opportunities in both tools and how we think and treat AI. Innovation in

AI for cybersecurity will have to continue to keep up with the threat and technologies.

Let's look at some of the trends for AI in cybersecurity:

- *Increased Use of AI by Threat Actors*: as we have discussed earlier in this chapter, AI is being incorporated in many new and exciting cybersecurity tools and platforms. That helps us defend and respond to attacks more quickly and more efficiently. It doesn't end there; the threat actors are doing the same. They are using AI in their tools. By using AI, they can incorporate automation in various attacks (e.g., malware). In some cases, the automation could allow an actor to develop malware that changes its tactics in real time. This could mean that it may bypass traditional defenses.

- *Emerging Technologies*: AI is going to be everywhere, not just cybersecurity. AI will play a role in emerging technologies such as quantum computing, internet of things (IOT) and other technologies. Quantum computing is a hot topic right now. Experts are saying that in five to 10 years quantum computing will be available and will be able to crack our current encryption algorithms in no time. The good news is that the National Institute of Standards and Technology (NIST) released the new encryption standard for quantum computing (NIST, "NIST Releases First 3 Finalized Post-Quantum Quantum Encryption Standards, n.d.). How does this impact AI? With this new technology, AI will play a big part in rethinking the cybersecurity strategy. IOT has been around for a few years, and it has had many issues, especially in the cybersecurity space. With AI in the forefront, being able to analyze data in real time could help improve the overall security of IOT.

- *Proactive Threat Detection*: machine learning has been around for a while and has been incorporated in various security-related tools, but with the advancement of AI, threat detection will advance. Threat detection will become even more automated and advanced with AI. In cybersecurity there is a process called threat hunting. This process is when cybersecurity professionals actively go out and try and find threats. There is a whole industry around threat hunting within cybersecurity. Now, with

AI advancing at the speed of light it's possible to automate this process. This means that our ability to detect the threat actors will increase.

- *Regulatory and Ethical Changes*: as AI becomes more prevalent in cybersecurity and in general, the need for frameworks, governance, and regulation increases. Recently the EU has passed its own AI regulation called the AI Act (Comission, n.d.) that is expected to shape AI all over the world. But that does not take care of the ethical concerns tide to AI. Some of the ethical considerations that will need to be addressed are as follows:

  o   Fairness and bias—ensure AI does not discriminate

  o   Transparency—how the AI system works

  o   Trustworthiness—building steps with users that ensure transparency

Conclusion

As we've seen, AI is already making waves in the world of cybersecurity, and this is just the beginning. The potential for AI to transform how we protect our systems and data is enormous, but let's not sugar-coat it—there are also risks. Just like with any new tool or technology, it's a double-edged sword. AI can be a game changer for defending against cyberattacks, but the same technology can be used by the "bad guys" to launch more sophisticated attacks. It's a race, and no one can afford to sit on the sidelines.

Think about it this way: AI in cybersecurity is kind of like putting a super-smart assistant in charge of monitoring everything. It can work faster than humans, analyze huge amounts of data, and pick up on patterns that we might miss. That's awesome. But just like any assistant, you still need to keep an eye on it. If the data it's learning from gets tampered with or it gets tricked, things can go wrong. The potential for AI-powered cyberattacks is very real, and threat actors are already leveraging these technologies to exploit vulnerabilities. That's why cybersecurity professionals need to stay one step ahead.

It's not just about playing defense. AI helps security teams move from reacting to threats to actively hunting for them before they become a problem. That's huge and it's something that's only going to get more important in the future. The more we rely on AI to help defend our networks and systems, the better our chances of catching these threats before they can do damage. But remember, there is no magic bullet in cybersecurity. AI is powerful, but it's not perfect.

And that brings us to the future. AI will continue to evolve, and with that evolution comes the need for a balance between man and machine. We're going to need more collaboration between humans and AI. While AI can handle the heavy lifting of processing data and identifying patterns, it's the human element—intuition, experience, and strategic thinking—that will ultimately make the difference. AI is an amazing tool, but we can't forget that it's just a tool. AI needs direction and oversight (they didn't learn that in *Terminator*), and that's where skilled cybersecurity professionals come in.

The future of AI in cybersecurity also brings up some interesting challenges when it comes to ethics and regulations. As AI becomes more widespread, we're going to see more calls for transparency, fairness, and security in how these systems operate. Companies and governments will need to ensure that the AI they use is ethical, secure, and doesn't infringe on privacy. It's a big ask, but it's also an essential one if we want to keep AI working for us—and not against us.

So, where does that leave us? AI is here to stay, and its role in cybersecurity will only grow from here. It's not a question of whether AI will change the game—it already is. The real question is, are we ready to harness it responsibly? The answer should be a resounding yes. By staying proactive, keeping the human element in the loop, and embracing AI's potential, we can build a more secure future. Now's the time to take what we've learned and make sure we're not just reacting to the threats of today but anticipating the challenges of tomorrow. Let's stay ahead of the curve.

References

https://www.ibm.com/think/topics/ai-security

https://techcrunch.com/2024/05/02/microsoft-bans-u-s-police-departments-azure-openai-facial-recognition/

https://www.nist.gov/news-events/news/2024/08/nist-releases-first-3-finalized-post-quantum-encryption-standards

About the Author

Toby Miller is an experienced chief information security officer (CISO) with a strong background in cybersecurity, risk management, and AI. Holding an MBA and industry certifications, Toby has extensive leadership expertise in developing and implementing security strategies to safeguard organizations from evolving cyber threats. His work spans across various sectors, where he's responsible for managing security frameworks, compliance, and risk assessments to mitigate potential breaches and ensure robust defenses.

As a thought leader, Toby actively contributes to the cybersecurity community by co-authoring books, writing on AI-related topics, and creating insightful content to educate both professionals and general audiences. With over 20 years of experience, he excels at simplifying complex topics, offering practical solutions, and empowering teams through collaboration and innovation. In addition, Toby is highly focused on continuous professional development, consistently enhancing his skills and staying abreast of the latest trends in security and technology.

Email: william.toby.miller@gmail.com

THE TRANSFORMATIVE IMPACT OF AI ON SOCIETY AND CYBERSECURITY

By Carolina Monge Palazón
Co-founder of Aittitude Consulting Tech, AI & ML
Madrid, Spain

*Artificial intelligence is not a substitute for human intelligence;
it's an amplifier of human ingenuity.*
—Fei-Fei Li

AI heralds a new era, similar to previous industrial revolutions, with the capacity to fundamentally alter the way we live and work. And just as happened in the mid-18th century with the Industrial Revolution, history repeats itself: a lack of knowledge makes us see monsters and dragons. Can the reader imagine what our civilization would be like if we had let ourselves be overcome by fear and listened to those who said the steam engine was an invention of the evil one? Here is a brief example of the advances that emerged from the first Industrial Revolution:

- The steam-powered automobile (1770)
- Watt's steam engine (1778)
- The mechanical loom (1785)
- Gas street lighting (1802)
- The railway (1814)
- The typewriter (1829)
- The telephone (1876)
- The gas-powered automobile (1886)

Galileo and Tesla were labeled madmen, Einstein as dangerously pre-communist, Darwin as a heretic ... And so many other brilliant minds have made it possible for us to be where we are today.

If we analyze the origin of this fear of the new, we will see that there are various factors:

Cultural Narratives: different cultures have different narratives about technology and AI. For example, Western cultures often depict AI in dystopian scenarios, influenced by films and literature that emphasize themes of rebellion and loss of control, such as *Terminator* and *Ex Machina*. In contrast, other cultures may view robots and AI more positively, as tools to improve harmony and efficiency in society.

Historical Context: from an anthropological perspective, fears of new technologies frequently reflect historical anxieties. The Industrial Revolution faced similar fears, as it was feared that innovations would disrupt social structures and displace jobs. However, history shows that societies adapt and find ways to integrate new technologies, ultimately benefiting from them.

Fear of the Unknown: human beings have an inherent fear of the unknown, which is amplified in the context of rapidly advancing technologies like AI. This fear can be exacerbated by a lack of understanding of how AI works and its implications for society. Many people anthropomorphize AI, attributing human-like qualities to machines, which can lead to irrational fears about their capabilities and intentions.

Projection of Social Flaws: the fears surrounding AI may also reflect deeper social problems, such as concerns about control, power dynamics, and ethical considerations. These fears can be seen as projections of society's flaws, where perceived threats from AI highlight existing vulnerabilities in human relationships and governance.

Education and Awareness: Increasing public understanding of AI and its mechanisms can help alleviate fears. By demystifying AI and promoting debates about its ethical use, societies can foster a more balanced view of technology as a tool rather than a threat.

It is understandable that despite these advances, the integration of AI raises concerns about job displacement, privacy, and cybersecurity. And this is where experts must provide guidance to raise awareness in society about the need, on the one hand, to overcome fear and embrace change, but on the other hand to do so responsibly and to bear in mind that it is a technology with exponential growth: it is and will be so in both good hands and in the hands of those who want to make illegitimate use of it. Like everything in life.

Let's briefly analyze these fears.

The fear of job loss is exaggerated. While AI may automate some jobs, it is also creating new ones that we can't even imagine yet, similar to how the internet led to jobs in social media and app development. By focusing on education and lifelong learning, we can prepare for these new opportunities.

AI itself is not evil, "out of control," or necessarily dangerous. What is dangerous is how we decide to use it. Every technology ever invented has been used for evil purposes, but the benefits have outweighed the risks over time.

Machine learning, at its core, is the process of giving a computer a large amount of data and an algorithm with the goal of the computer becoming more efficient at the algorithm through repetition. In reality, it does not acquire intelligence or consciousness. In fact, the idea of an uncontrollable superintelligent AI is still speculative and far from our current reality. Many scientists are already working on safeguards and preventive measures. We should assume that this technology, like all others ever invented, will be tamed and improved over time. The fears surrounding AI are not insurmountable obstacles, but rather guidelines that lead us to a more mindful and responsible use of technology.

AI is a tool created by humans, and just as we have learned to understand complex systems like our bodies and nature, we will also learn to understand AI.

It is imperative to establish strong ethical guidelines and regulations for the development and implementation of AI to prevent its misuse. Ultimately, a global collaborative effort is essential to create a secure digital landscape where AI serves as a force for good.

Artificial intelligence (AI) is rapidly transforming societies, economies, and industries worldwide. As this technology becomes increasingly integrated into our daily lives, it is crucial to understand its profound impact on civilization and the potential risks associated with its deployment without adequate professional oversight. This chapter explores how AI is revolutionizing various sectors, the challenges posed by unregulated AI commercialization, and the importance of responsible governance.

A Double-Edged Sword: Navigating the AI Revolution

AI is undoubtedly ushering in a new era, reminiscent of past industrial revolutions, with the potential to fundamentally transform the way we live and work. This technological behemoth is reshaping industries from healthcare to finance.

In medicine, for example, it is accelerating drug discovery, enabling precision medicine, and improving treatment outcomes for patients. Similarly, the financial sector is undergoing a metamorphosis, with AI optimizing trading algorithms and enhancing fraud detection.

However, this digital revolution is not without its challenges. Job displacement looms as a significant concern as AI automates routine tasks, potentially altering the employment landscape. Additionally, the digital divide presents a critical issue, with unequal access to AI benefits across different regions and demographic groups. Moreover, as AI systems increasingly rely on vast amounts of personal information, concerns about data privacy and security intensify.

Striking the right balance between harnessing AI's potential and mitigating its risks is paramount. Governments, businesses, and individuals must collaborate to ensure that this powerful technology is used ethically and responsibly for the benefit of all. Ultimately, the success of the AI era depends on our ability to adapt, improve our skills, and leverage human ingenuity to complement AI's capabilities. By doing so, we can navigate the complexities of AI and build a future that leverages its transformative potential for the greater good.

Artificial intelligence has emerged as a formidable asset in bolstering cybersecurity defenses. These sophisticated AI systems demonstrate remarkable prowess in processing immense volumes of data, enabling real-time threat detection and response. Consequently, this technological advancement has led to a significant enhancement in both threat identification capabilities and incident response protocols. Some key advantages of AI in cybersecurity are remarkable, and we can see deep improvements in very specific fields like threat detection.

AI-powered systems have significantly enhanced the strength of solutions by analyzing network traffic patterns and user behaviors to identify anomalies that may indicate cyberattacks. Machine learning algorithms can detect subtle patterns and evolving threats that traditional rule-based systems may miss, enabling more proactive and accurate threat detection.

Advanced Pattern Recognition, Behavioral Analysis and Real-Time Monitoring, Threat Intelligence Integration, and Some Other Superpowers for Cybersecurity Teams

As already mentioned, machine learning algorithms analyze vast amounts of network traffic data to identify subtle patterns and anomalies that indicate threats. Unlike traditional rule-based systems AI can detect emerging and previously unknown attack patterns.

Behavioral analysis wise, AI systems establish baselines of normal user and network behavior. Any deviations from these baselines can be flagged as potential threats, enabling the detection of insider threats and compromised accounts.

Real-time monitoring has been a game changer in some industries like banking, betting, and so on, as the analyses are not forensics anymore. Responding immediately allows for immediate loss prevention as AI enables continuous, real-time analysis of network activity. Threats can be identified and responded to immediately, rather than after the fact.

AI systems fueled with historical data and trends can predict potential future threats and vulnerabilities with predictive analytics, allowing organizations to take preventive measures and strengthen defenses proactively. It also gives the capability to rapidly incorporate new threat intelligence data to improve detection of emerging attack vectors, which, at the same time, reduces false alarms. This allows analysts to focus on genuine threats and security teams to focus on the most critical issues first.

By leveraging these AI capabilities, organizations can detect threats faster, more accurately, and at a larger scale than traditional methods. This proactive approach significantly enhances overall cybersecurity postures and resilience against evolving threats.

Rapid Threat Containment, Streamlined Workflow Automation, and Intelligent Triage and Prioritization, and Some Other Superpowers for Enterprises

Thanks to this technology, we now can automatically execute containment actions as soon as a threat is detected, isolating affected systems or network segments to prevent lateral movement such as blocking malicious IP addresses or terminating suspicious processes, quarantining infected files, or disabling compromised user accounts. This rapid automated response significantly reduces response times and potential damage, losses, and fines.

Human-in-the-Loop Approach

This adaptive approach allows the incident response capability to evolve alongside the threat landscape. However, it's important to maintain human oversight and incorporate "human-in-the-loop" approaches for critical decision-making.

The human-in-the loop approach is a powerful source to identify areas for improvement in processes and tools, to measure and optimize key metrics like mean time to detect (MTTD) and mean time to respond (MTTR), and to justify security investments to leadership. But it is crucial to valuable insights, and it's crucial to maintain human oversight.

We can conclude that artificial intelligence (AI) is a transformative technology with the potential to revolutionize society and industries. While its integration presents challenges, such as job displacement and privacy concerns, the benefits are significant. AI enhances efficiency, improves decision-making, and drives innovation in various sectors.

In cybersecurity, AI plays a crucial role in detecting and responding to threats. By analyzing vast amounts of data, AI can identify anomalies and patterns that indicate potential attacks. Additionally, AI can help establish baselines for normal user and network behavior, flagging any deviations as potential threats. This proactive approach can significantly enhance cybersecurity postures and resilience.

However, it's essential to approach AI with a balanced perspective. While AI offers immense potential, it's crucial to address concerns about misuse and ensure ethical development and deployment. By establishing robust regulations, promoting transparency, and fostering collaboration among stakeholders, we can harness AI's benefits while mitigating its risks.

Ultimately, the success of the AI era depends on our ability to adapt, upskill, and leverage human ingenuity to complement AI's capabilities. By doing so, we can navigate the complexities of AI and build a future that benefits humanity as a whole.

Final Thoughts

The question of whether a computer can think is no more interesting than the question of whether a submarine can swim.
— Edsger W. Dijkstra

We find ourselves at the dawn of a new technological epoch, where the confluence of artificial intelligence and cybersecurity unfolds a tapestry of unparalleled prospects and daunting trials. This intricate interplay between AI and cybersecurity transcends mere passing fashion; it heralds a seismic shift in our approach to digital safeguarding and risk management.

The infusion of AI into cybersecurity frameworks ushers in an era of dynamic defense systems, endowed with the remarkable capacity to evolve instantaneously, countering the ever-increasing sophistication of cyber threats. Yet, this technological symbiosis also beckons us to confront profound quandaries concerning privacy, ethics, and the essence of human-machine collaboration in fortifying our digital realms.

As we traverse this labyrinthine landscape, it becomes incumbent upon policymakers, technologists, and society writ large to engage in nuanced discourse and concerted endeavors. Our collective aspiration must be to harness AI's transformative potential in cybersecurity while concurrently addressing its latent pitfalls, ensuring that our quest for digital fortification does not exact a toll on individual freedoms or ethical principles.

In essence, this amalgamation of AI and cybersecurity represents not merely a technological leap, but a watershed moment in our digital odyssey. It compels us to reimagine our security paradigms, to innovate with a sense of responsibility, and to craft a future where technology stands as a bulwark, rather than a weapon, in the perpetual struggle for digital integrity and resilience.

This contemplation encapsulates the cardinal themes of AI and cybersecurity, offering a forward-looking vista and a reflective denouement that invites deeper rumination on this pivotal subject.

References

https://www.leewayhertz.com/ai-in-incident-response/

https://www.techtarget.com/searchsecurity/tip/Incident-response-automation-What-it-is-and-how-it-works

https://atos.net/en/lp/detect-early-respond-swiftly/ai-powered-incident-response-harnessing-the-potential-of-self-healing-endpoints

https://www.pagerduty.com/resources/learn/what-is-incident-response-automation/

https://www.rezolve.ai/blog/automated-incident-response-everything-you-need-to-know

https://blog.barracuda.com/2024/07/01/5-ways-ai-is-being-used-to-improve-security--automated-and-augme

https://www.blinkops.com/blog/the-top-5-ways-ai-is-automating-cybersecurity-incident-response

https://www.cyberdefensemagazine.com/ai-and-cybersecurity-mitigating-risks-and-safeguarding-digital-assets/

About the Author

Carolina Monge Palazón is a prominent professional in the field of technology and cybersecurity, with over 15 years of experience in the industry. She currently serves as the executive director of Aittitude Consulting Tech. Throughout her career, Carolina has held key roles such as Alliance Manager and Strategy Director in well-known corporations. Her expertise focuses on international market expansion, with a specialization in Artificial Intelligence, Machine Learning, and Cybersecurity.

Carolina is distinguished by her innovative vision and commitment to continuous education, which allows her to stay at the forefront of the latest trends and advancements in her field. Her approach is characterized by a pursuit of impartiality and neutrality in consulting operations, avoiding commercial ties with specific vendors.

Carolina lives happily in Madrid and enjoys nature walks whenever she can escape from so much technology.

Email: c.m.p@aittitudeconsulting.com
LinkedIn: https://es.linkedin.com/in/carolina-monge-palazon-5423ab20

AI FOR SOCIAL IMPACT: ADVANCING GENDER EQUALITY AND STEM EDUCATION IN AFRICA

By Gloriana J. Monko
PhD Researcher; AI Expert, Inclusion in AI
Tokyo, Japan | Dodoma, Tanzania

We will not only use the machines for their intelligence,
we will also collaborate with them in ways that we cannot
even imagine.
—Fei-Fei Li

Artificial intelligence (AI) has revolutionized various sectors, from healthcare to finance, by providing innovative solutions to complex problems. Its potential for social impact is equally significant, particularly in addressing deep-rooted issues like gender inequality and enhancing educational opportunities. This chapter explores the transformative power of AI in advancing gender equality and promoting STEM (science, technology, engineering, and mathematics) education in Africa. We aim to highlight how AI can contribute to a more equitable and prosperous society by examining current challenges, thriving initiatives, and future possibilities.

The transformative potential of AI in addressing societal issues is gaining momentum globally, but its application in Africa carries unique significance. The continent faces many challenges, including gender disparities, limited access to quality education, and economic inequalities. With 70% of sub-Saharan Africa's population under the age of 30, the need to equip young people with relevant skills is urgent. AI can play a pivotal role in bridging these gaps by offering scalable solutions catering to African societies' unique needs. Moreover, integrating AI into social impact initiatives provides new opportunities to address gender-specific challenges, ensuring that women and girls are not left behind in the digital revolution.

Current State of Gender Equality in Africa

Gender equality in Africa has seen varying degrees of progress, with significant disparities still present across the continent. Women and girls often face challenges in accessing education, healthcare, and economic opportunities due to cultural norms, economic barriers, and political instability. According to the latest data from the World Bank, 27% of seats in national parliaments in sub-Saharan Africa are held by women, slightly above the world average of 26.7%. However, this representation still reflects the broader gender disparity in leadership and decision-making roles. Furthermore, women in Africa are more likely to be engaged in informal and low-wage employment, with limited access to financial resources and land ownership. For example, women account for over 70% of the agricultural labor force in sub-Saharan Africa but control less than 20% of the land. These systemic inequalities hinder social and economic development, underscoring the need for targeted interventions.

The barriers to gender equality in Africa are multifaceted. Despite Nigeria being Africa's largest economy, only 6% of women own land, a factor perpetuating economic dependency. Additionally, gender-based violence remains a pervasive issue, with 45% of women in sub-Saharan Africa experiencing physical or sexual violence in their lifetime. Addressing these challenges requires a holistic approach that includes legal reforms, community education, and women's empowerment through access to resources and opportunities.

AI has immense potential to facilitate new economic opportunities for African women, mainly by providing tools and resources to enhance their productivity and entrepreneurship. By harnessing AI's power, we can ensure greater access to education, improve women's economic empowerment, tackle gender-based violence, and promote gender equality in leadership, accelerating social and economic development across the continent.

STEM Education in Africa

STEM education is critical for fostering innovation and economic growth. However, in many African countries, STEM education faces numerous challenges, including inadequate infrastructure, a lack of trained teachers, and insufficient learning materials. For girls and women, these barriers are compounded by societal expectations and gender biases that discourage their participation in STEM fields. According to UNESCO and the World Bank, only 30% of African science and engineering researchers are women. This gender gap in STEM education limits the region's potential for technological advancements and economic diversification.

The challenges in STEM education are evident in both primary and tertiary levels of education. In countries like Tanzania, the lower participation of female students in science subjects is primarily due to the lack of adequate facilities, a shortage of female role models, and insufficient encouragement. These barriers prevent many girls from pursuing careers in STEM, contributing to the gender disparity in these fields. Furthermore, the curriculums in many African countries are outdated, failing to incorporate the latest technological advancements or align with the demands of the modern workforce. This disconnect between education and industry needs exacerbates unemployment rates among youth, particularly young women, who are underrepresented in STEM careers.

Addressing these challenges requires a multifaceted approach. Programs such as TechWomen, which connects emerging women leaders in STEM with their counterparts in Silicon Valley, provide mentorship and exposure to cutting-edge technologies. Additionally, initiatives like the African Institute for Mathematical Sciences (AIMS) offer scholarships and advanced training for women in mathematics, aiming to create a pipeline of female scientists and engineers who can contribute to Africa's development.

Role of AI in Promoting Gender Equality

AI can be a powerful tool in addressing gender inequalities by providing data-driven insights and scalable solutions. Part of the AI Mindset should be AI-powered platforms that can analyze large datasets to identify gender biases in hiring practices, educational opportunities, and resource allocation. For instance, an AI-driven tool using machine learning algorithms can predict and address dropout rates among girls in schools and help to keep more girls in education. Countries like Rwanda have been at the forefront of using AI and data analytics in various sectors, including education, as part of their broader strategy to integrate AI into national policies. Additionally, AI can enhance women's healthcare access by providing remote diagnostics and personalized treatment plans, addressing the gender disparity in healthcare access.

One of the key advantages of AI is its ability to process and analyze large amounts of data quickly, uncovering patterns that might be missed by human analysis. For example, in employment, AI can analyze job postings and hiring data to identify gender biases, such as the use of gendered language that may discourage women from applying for certain positions. AI can also be used to design interventions that specifically target gender disparities. For instance, AI algorithms can be employed to monitor and ensure equal pay for equal work, a persistent issue in many African countries.

Moreover, AI can support the development of gender-sensitive policies by providing insights into how men and women are affected by various issues, from climate change to economic policies. In healthcare, AI-driven tools can be used to track and address maternal health challenges, a critical area of concern in Africa, where maternal mortality rates remain high. By providing real-time data and predictive analytics, AI can help healthcare providers offer more timely and targeted interventions.

AI in Enhancing STEM Education

AI has the potential to transform STEM education by making learning more personalized, interactive, and accessible. AI-driven educational platforms can adapt to individual learning styles and paces, providing customized content and feedback. For example, M-Shule, an AI-powered mobile learning platform in Kenya, uses SMS-based lessons to reach students in remote areas, improving their math and literacy skills. Another recent initiative by Elimunity called AI-STEM is at the forefront of transforming STEM education through cutting-edge AI technology in Tanzania. This platform leverages AI to bridge the gap between traditional teaching methods and the demands of modern education, ensuring that every student can excel in STEM subjects. AI can also facilitate virtual labs and simulations, allowing students to conduct experiments and gain practical experience without the need for expensive physical infrastructure. These innovations can make STEM education more inclusive and engaging for girls and women.

In addition to personalized learning, AI can help address the shortage of qualified STEM teachers in many African countries. AI-driven tools can offer real-time support to teachers, providing lesson plans, grading assistance, and even professional development resources. This support can be particularly beneficial in rural areas where access to quality education is limited. Furthermore, AI can help bridge the language barrier that often hinders learning in multilingual countries. This can be achieved by providing content in local languages, tailoring instruction to the linguistic needs of students, and making STEM education more accessible to a broader population.

AI is also opening new avenues for practical, hands-on learning. Students can explore complex scientific concepts with greater immersion and engagement through virtual reality (VR) and augmented reality (AR). For example, students can virtually dissect a frog or explore the surface of Mars, experiences that would be impossible in a traditional classroom setting. These technologies can inspire more girls to pursue STEM by making science and technology more relatable and exciting.

Challenges and Ethical Considerations

Implementing AI for social impact is not without challenges. Data privacy, algorithmic bias, and the digital divide must be carefully addressed to ensure that AI applications are fair and equitable. Algorithmic bias can perpetuate existing inequalities if not properly managed. For example, AI systems trained on biased datasets may reinforce gender stereotypes. To mitigate these risks, it is crucial to develop transparent and accountable AI systems, involve diverse stakeholders in the design process, and implement robust data governance frameworks. Additionally, efforts must be made to bridge the digital divide by ensuring marginalized communities have access to the necessary technology and digital literacy skills.

The ethical considerations surrounding AI are critical in the African context, where issues of inequality and access are already pronounced. For example, the use of AI in hiring processes could inadvertently perpetuate gender biases if the training data reflects existing disparities. Similarly, AI-driven educational tools must be designed to accommodate diverse learning needs and avoid reinforcing stereotypes. It is also essential to ensure that AI does not exacerbate the digital divide, leaving rural or economically disadvantaged communities further behind. This requires concerted efforts to expand access to digital infrastructure and ensure all individuals have the skills necessary to participate in the digital economy.

Furthermore, clear regulatory frameworks are needed to govern the use of AI, particularly in sensitive areas such as healthcare and education. These frameworks should ensure that AI applications are used ethically and that the rights of individuals, particularly those from marginalized communities, are protected. Collaboration between governments, civil society, and the private sector is crucial in developing these frameworks and ensuring they are implemented effectively.

Policy and Institutional Support

Government policies are essential for creating an environment where AI can thrive and generate social impact. Policies that promote gender equality, support STEM education, and encourage innovation are particularly critical. Countries like Ghana, Rwanda, Egypt, and South Africa have implemented national strategies to integrate AI and digital technologies into their educational systems. These national efforts help lay the foundation for more inclusive and innovative AI applications across the continent.

Policy initiatives promoting gender equality in AI and STEM are also significant. For instance, Rwanda has introduced policies that require gender parity in STEM-related scholarships and training programs. Such initiatives encourage more women to pursue STEM careers and ensure that AI solutions are developed with a gender-sensitive lens. Additionally, regional collaborations like the African Union's Agenda 2063 emphasize the importance of science, technology, and innovation for Africa's development, providing a framework for countries to align their AI and STEM education policies.

Institutions also play a pivotal role by funding AI research, supporting public-private partnerships, and fostering collaborations between academia, industry, and civil society. Universities and research centers, in particular, are crucial in advancing AI for social impact. By integrating AI into their curricula and research agendas, these institutions contribute to developing AI tools that address local challenges. These efforts help establish a supportive ecosystem for AI-driven initiatives that advance gender equality and STEM education. A notable example is the establishment of the multidisciplinary Anglophone AI4D Lab in Tanzania, which promotes capacity development, research, and innovation in responsible AI while championing gender equality and inclusion. Moreover, through partnerships with industry and government, academic institutions can bridge the gap between research and practical applications, ensuring that AI solutions are innovative and relevant to African societies' unique needs.

Future Directions

The future of AI in promoting gender equality and STEM education in Africa is promising. Emerging trends like AI-powered mentorship programs, online learning platforms, and data-driven policy-making hold great potential. AI can also play a role in addressing broader societal challenges, such as climate change and healthcare, by providing innovative solutions that benefit all members of society. Continuous investment in AI research, infrastructure, and human capital is necessary to realize this potential. By fostering a culture of innovation and inclusivity, Africa can harness the power of AI to drive social and economic transformation.

As AI technology evolves, new opportunities will emerge to leverage AI for social good. For example, AI-driven platforms could provide real-time data on gender disparities in various sectors, allowing governments and organizations to make more informed decisions. Additionally, AI could be used to develop targeted interventions for specific groups, such as women entrepreneurs or girls in rural areas, helping address their unique challenges. Furthermore, integrating AI with other emerging technologies, such as blockchain and IoT (internet of things), could create new avenues for social impact, from ensuring fair trade practices to improving access to clean water and sanitation.

Looking ahead, it is crucial to ensure that the benefits of AI are equitably distributed across society. This will require ongoing efforts to promote digital literacy, particularly among women and girls, and to ensure that AI tools are designed and implemented in an inclusive and accessible way. By prioritizing gender equality and STEM education in the development and deployment of AI, Africa can position itself as a leader in the global AI landscape, driving innovation and social progress.

Conclusion

Integrating AI into social impact initiatives in Africa presents a unique opportunity to address some of the continent's most pressing challenges. By leveraging AI's capabilities, African countries can make significant strides toward achieving gender equality, improving access to quality education, and fostering economic development. However, this will require a concerted effort from all stakeholders, including governments, the private sector, academia, and civil society. With the right policies, investments, and collaborations, AI can catalyze social change, helping create a more just and inclusive society.

References

UNESCO Institute for Statistics. (2021). Women in Science. Retrieved from https://uis.unesco.org/en/topic/women-science

World Bank. (2023). Gender Data Portal. Retrieved from https://genderdata.worldbank.org/en/regions/sub-saharan-africa#:~:text=27%25%20of%20seats%20in%20national,the%20world%20average%20of%2026.7%25

Food and Agriculture Organization of the United Nations (FAO). (2023). Women in Agriculture: Closing the Gender Gap for Development. Retrieved from https://www.fao.org/gender/learning-center/thematic-areas/gender-equality-and-women-empowerment/

African Union. (2020). Agenda 2063: The Africa We Want. Retrieved from https://au.int/en/agenda2063/overview

M-Shule. (2023). Empowering Africa's Future with Mobile Learning. Retrieved from https://www.mshule.com/

African Institute for Mathematical Sciences (AIMS). (2022). Women in STEM. Retrieved from https://nexteinstein.org/aimswis/

TechWomen. (2021). Connecting Women in Tech for Global Impact. Retrieved from https://www.techwomen.org/

World Economic Forum. (2018). Women farmers are the backbone of Africa's agricultural workforce, but policy changes are needed. Retrieved from https://www.weforum.org/agenda/2018/03/women-farmers-food-production-land-rights/

United Nations Development Programme. (2020). Gender Equality Strategy. Retrieved from https://www.undp.org/publications/undp-gender-equality-strategy-2020-annual-report

World Bank. (2021). *Nurturing Africa's next generation of female scientists*. Retrieved from https://blogs.worldbank.org/en/nasikiliza/nurturing-africas-next-generation-female-scientists

About the Author

Gloriana Monko is a Tanzanian AI expert specializing in natural language processing and is currently pursuing her PhD at SIT in Tokyo. Her expansive research interests span machine learning, NLP, computer science, and initiatives in STEM, gender equality, and transformative learning.

Gloriana has led and collaborated on impactful projects, including transforming employability for social change in East Africa, designing low-cost physics toolkits to boost girls' Participation in STEM, and establishing a multidisciplinary gender-inclusive AI research lab. She co-founded Elimunity and Women Supporting Women in the Sciences (WS2) and founded EmpowerHer Mind.

Gloriana has been a keynote speaker and presenter at numerous conferences, winning awards like the best-presented papers at international conferences. She has published on AI algorithms, STEM education, and computer science. She also serves as liaison manager and gender coordinator at the Anglophone AI4D lab, further exemplifying her commitment to advancing gender equality and inclusivity in science and technology.

Email: gmonko24@gmail.com
Websites: https://www.elimuunity.com/
https://www.empowerhermind.org/

CHAPTER 22

UNMATCHED PRODUCTIVITY

By Christopher Narowski
AI Strategy Consultant & CEO of Dumbify
Colorado Springs, Colorado

The better we get at getting better, the faster we will get better.
—Douglas Engelbart

"I strive to be lazy. No really, *I love being lazy!* If I find any tasks or processes that are repetitive and that can be automated, I'll find a way to automate them."

These were the exact words I used when applying to my former role at Meta. While my phrasing might not have been ideal (and my interviewer made sure to tell me, after chuckling, that I should phrase that differently), it perfectly captured my intentions for that role. I've always been focused on getting as much done in as little time as possible while ensuring the quality of my work is still the best of what I can do.

It wasn't really laziness either, but it was about removing even the smallest tasks that annoyed me, so I could focus on the things I really enjoyed. Luckily, not long after joining Meta, ChatGPT was released, and it led to a new level of efficiency I couldn't have dreamed of in my previous roles. Driven by my strong desire to be "lazy," I dove deep into AI, searching for ways to approach streamlined process development. Together with my team (some of the smartest and kindest people I've ever met), we used it to transform everything, from generating data transformation scripts, to solving problems that normally took hours to debug, and even quickly generating ideas for responding to difficult or upset clients.

My journey took another leap forward once open source projects like Llama, Qwen, and Stable Diffusion became available. These tools allowed me to automate processes, offer improved consultation services, and optimize business workflows while still being able to work on the things I love. I've been incredibly fortunate to explore the intersection of AI and productivity first hand, and my goal here is to share that knowledge, helping others stay competitive in today's hyper-efficiency focused world.

A New Standard to Efficiency (A Data-Driven Approach)

Now, I want to avoid providing any information without really diving into proof of just how transformative AI can be. Inversely, I also want to avoid being mind-numbingly boring with data on how important AI is to your success and efficiency. So I will aim to provide a mix of both as we go through this, taking both anecdotal personal experience and mixing it with studies that are publicly available.

At Meta, I was in charge of a support engineering team that handled everything from easy-to-answer questions, to diving deep into mobile SDKs, REST APIs, or just making things work in highly customizable asynchronous environments where concurrent tasks could overlap each other (race conditions). When AI was introduced to us, there was a bit of apprehension toward adopting it in our day-to-day work as there was a general fear of it replacing what we do; however, after exploring it deeper we found it best viewed as a strong tool for experts, and an even better tool for onboarding and filling in knowledge gaps. We were able to drastically improve our response times, handle times, and even our deflection rate, giving us more room to further improve the processes we had and focus on skill growth.

This directly relates to a study from the National Bureau of Economic Research where they found the average agent would see a 14% increase in productivity, and novice/entry-level workers saw a 34% improvement (Brynjolfsson et al., 2023). This improvement here was measured in how many issues per hour an agent was able to navigate. What might be most interesting in this study is that it found expert workers did not see much of an improvement. I strongly believe this can be summed up to how AI is utilized and then how it is measured in terms of productivity. For our team, I did find that that was indeed true in terms of total number of completed items, but the difficulty of the tasks my team was able to take on was the major kicker. Rather than getting stuck on deeper SDK or coding issues that we would normally send to our engineering team, we were instead able to tackle more of these difficult issues. This meant our team was directly impacting in a positive way, what our software engineering team focused on, and we were able to start seeing improvement company wide. The use of AI for our team quickly transitioned from being a nice idea, to a necessity without requiring much more than use of a pre-built tool. Unfortunately, it isn't always that easy for everyone.

The Theory of Overcoming Hurdles Faster with AI

While I can't speak to the metrics our team might have benefited the development team, I can certainly speak to how our collaboration changed. It was fast paced! We were able to collaborate quickly on issues and talk to a certain level of expertise our engineering team was not generally used to. This came down to a new automation we had in place that we didn't even consider being an automation.

Before I dive more into that, I want to mention how there's generally a direct relationship between fast results and the quality of work. If you increase how fast you handle something, it generally means quality suffers, and vice versa. This is really where AI shines!

While AI might not change the problems you encounter, it will change how fast you can find solutions to those problems. Maintaining accuracy while increasing speed is the "holy grail" of any productivity-driven workflow. I can say from experience, most of the time spent on a problem is either spent working through trial-and-error workflows or spending hours digging on Google, Reddit, Stackoverflow, or any other community-driven forum. This is the process that needs the most refinement, and up until now, it isn't something I would have considered to "automate."

While our team mostly consisted of developers despite being a support team, there was certainly a knowledge gap between our platform use experts and our platform code experts. The efficiency we gained in that communication and collaboration led to happier developers, faster resolution of internal tasks, and an overall sense of comradery that we didn't have before. While AI didn't eliminate our difficult problems, it was an invaluable tool in the process of finding, implementing, and communicating solutions quickly.

Now outside my anecdotal and maybe overly positive outlook on AI's use, it can also be directly seen in a study from MIT where programmers using CoPilot saw an improvement of being 55.8% faster (this translates to a whopping 126% increase in throughput), without a change in the ability to complete the assignment. This sounds like some fairly amazing metrics, but it isn't as easy as just providing users with tools. Just like any automation tool put in place, there has to be a process to not only ensure adoption, but to measure its benefit.

Introducing the Double Diamond Design Process with AI

I was first introduced to the Double Diamond Design Process in an AI design course, and I quickly realized it can be applied to far more than just design or programming. It is a powerful approach to solving a wide range of challenges in both personal and business workflows. This process consists of two segments (diamonds), discovering and defining the problem, followed by developing and delivering the solution. While it sounds like it might be complex, I assure you it creates a simplified approach that may come across as just being common sense.

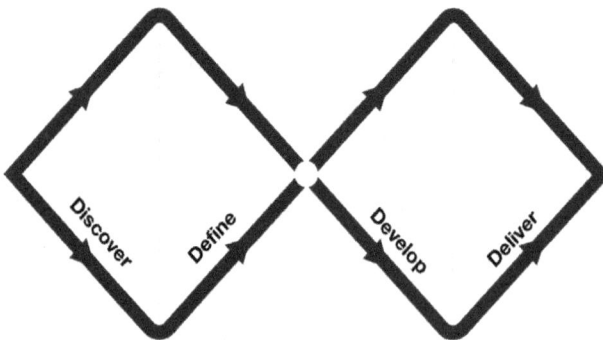

(The Double Diamond by the Design Council)

Discover: Understand the Problem in Detail

In this first phase, you take time to discover all aspects of a problem. This might mean gathering data, looking into existing workflows, or speaking to stakeholders to uncover inefficiencies or bottlenecks. If it is more individually focused, then take the time to really break down some of your small annoyances you have with regularly completed tasks.

Hopefully, by the end of this stage, it should feel like you've had a good venting session and you've identified issues you might not have realized were there in the first place. A great goal checklist might include:

- *Data Collection*: be sure to consider workflows, processes, and areas where most of your time is being spent.

- *Feedback*: if you're working with teams to do this, then engage with your team to identify pain points they encounter daily, especially in terms of repetitive tasks.

- *Performance*: get a snapshot of current metrics. This is extremely important since you need to identify what you are starting with to measure success.

The goal from this phase is to identify almost a wish list of things you would like to be (or might already be) lazy about.

Define: Refining the Problem into Actionable Objectives

Once you have finished your discovery, you need to refine these areas into clear, actionable problems:

- Map out specific pain points such as slow response times or regularly receiving bad data.

- Set measurable goals such as reducing response times by 20% or cutting data entry timelines in half.

- Determine the types of problems you encounter. Is there a theme in repetitive tasks, do you have poor data analysis, or do you have low customer satisfaction?

Here you should clearly know what your problems are and hopefully start to see ideas on what can be done to address them.

Develop: Creating AI-Driven Solutions

Now we get to the fun part, which is creating solutions. Let me start by saying this: telling people to just use AI will not work. I've seen this time and time again where organizations purchase great tools, provide a login, and then expect teams to figure it out with zero direction. You must develop full mindset and processes around how these tools can be implemented:

- Prototype solutions and make sure the tool in use matches the intended use case, whether it's a chatbot to reduce response times or some great AI analysis tool. There is a vast selection of tools to use, and it will take time to find out the best tool for the job.

- Ensure there is training to use the tools effectively. It often takes some research to know how to "work" with AI tools and prompt them effectively for the solutions you're looking for.

Deliver: Implement and Iterate

This is where you get to see all your hard work pay off. Rolling out your AI tool isn't just a one-time thing though, it's a never-ending process:

- Start with small group testing, especially if costs are involved. Select a small team, or if you are doing this solo, apply it to a subsection of your work.

- Gather feedback and track metrics. Compare it to the metrics you defined before and see if there is measurable improvement.

- Then finally, improve or change direction. Based on the performance of your solution, there might be a better tool to try.

Enter the AI-Powered Autodidact

Now that we've explored the Double Diamond framework, it hopefully hints that the success of AI in any workflow doesn't just depend on the tool we use but how well we implement it and the knowledge behind that effort. This brings me to my favorite concept: the AI-powered autodidact, an individual who leverages AI not just for productivity, but for self-learning.

The only way to improve these tools and processes further is to constantly and consistently be in a growth-centered mindset. Freeing up time with automation provides an opportunity to focus on the next challenge, or further growth. This naturally requires breaking down knowledge barriers. This quickly enables us not only to build better automations, but also sets us up in such a way to consistently learn on the job!

A great personal example I have is since we had more time to take on tougher issues, my team started encountering more issues with RegEx (regular expressions) patterns. Without getting into too much detail, RegEx is something you can use to quickly match data. Even for experienced developers, this can be challenging if you do not regularly work with them. So, in an effort to help my team in learning, I created a specialized GPT. This was built on top of ChatGPT, and it acted as a grumpy professor that gave you 20 randomized questions on RegEx patterns. If you got an answer wrong, it would chastise you (humorously) and then explain why it was wrong and provide reading on how to improve. While it was fun, it ended up being engaging and a great learning tool for my team.

All of this is to say, AI, when utilized effectively, can be a great tutor and resource for learning skills you might otherwise think are out of reach. It is a tutor that is always available, can be requested to talk in preferred style, will not be annoyed to repeat itself, and is available 24/7 with access to an unprecedented amount of data. If anything through this chapter, I hope you can consider the empowering nature of AI as a learning companion not only for your current role, but also in areas you might not have thought about. As I see it, this is the AI Mindset.

Thriving in the AI Era

To be clear, AI isn't just a magic button you press and—*voila!*—all your problems are solved, but it is setting a new precedent of productivity that almost necessitates you to embrace the change. That does come with a bit of caution or discretion, because although AI is great for automation and productivity, it absolutely needs oversight and refinement. If you just toss a question to ChatGPT or Claude.ai, and start copy/pasting that as responses to clients, you (and your client) are likely to be in for a shock.

I also want to call out that while most of my discussion has been specific to the tech industry, it isn't exclusive. A study on the productivity in manufacturing from 2010 to 2021 from Beijing Jiaotong University found that every 1% increase in the involvement of AI in business directly improved business resource efficiency by over 14%. So whether you're in tech, manufacturing, marketing, or even fitness, AI is transforming workflows across the board.

These tools are also only going to become more advanced, likely leaving a large gap between those that use AI tools and those that don't. While I don't suspect that AI will replace us, it will make us better at what we do, and you have to stay in the loop, continuously focus on learning, self-development, and sharpening your AI skills.

My best recommendation going into the AI era is to scan AI news and trends weekly, but please don't let it overwhelm you. Become familiar with the tools that make sense for you, or, even better, find areas that are exciting, and are something you would enjoy to learn more about. Try out AI tools, even if it's just for fun, to get a feel for what they can do. There are plenty of open source tools out there and even free lessons to learn how to build your own free version of popular tools. Dive into AI communities online and see how others are using it. You may find cool solutions you can build into your own workflows.

Lastly, I want to stress that AI isn't just about solving problems faster, it is a tool to empower you to focus more on what you love and excel at. So take some time to build great solutions and then enjoy some of the hard-earned opportunities to be "lazy" and explore the opportunities for growth you've put in place.

References

Brynjolfsson, Erik; Li, Danielle; Raymond, Lindsey R.: *Generative AI at Work*. Working Paper Series. National Bureau of Economic Research, April 2023, No. 31161. http://www.nber.org/papers/w31161

Gao, X.; Feng, H. AI-Driven Productivity Gains: Artificial Intelligence and Firm Productivity. *Sustainability* 2023, *15*, 8934. https://doi.org/10.3390/su15118934

Design Council. (n.d.). *What is the framework for innovation? Design Council's evolved Double Diamond*. Retrieved September 2, 2024 from https://www.designcouncil.org.uk/our-resources/the-double-diamond/

Peng, Sida; Kalliamvakou, Eirini; Cihon, Peter; Demirer, Mert. *The Impact of AI on Developer Productivity: Evidence from GitHub Copilot.* 2023. https://arxiv.org/abs/2302.06590

About the Author

Christopher Narowski has over 14 years of experience in SaaS, PaaS, and on-premise automation solutions, with a particular focus on AI, growth, and productivity enhancement. His passion for helping teams work smarter stems from his continuously growing interest in tech-driven efficiency. In addition to the tech industry, he has been a trainer and coach for gyms for over a decade, and this coaching mindset has helped him drive teams to hyper-growth and strong skill development.

At the time of this book, Christopher is the manager of technical account management at Kustomer, the CTO for Soteria Mobile, and is the founder of Dumbify, an AI-powered calendar productivity tool designed to help individuals and teams optimize their time. Dumbify, set to launch in 2025, aims to revolutionize how people manage their schedule, achieve goals, and improve their work-life balance.

Driven by a strong desire to help others grow, both personally and professionally, Christopher is dedicated to guiding individuals and organizations toward unlocking their full productivity potential.

Email: chris@ensif.com
LinkedIn: https://www.linkedin.com/in/cnarowski/

POWER OF DESIGN ANTHROPOLOGY: EMBEDDING SOCIAL CONTEXT FOR MEANINGFUL AI INTERACTIONS

By Ira Aurora Paavola
GenAI Lead, Business Anthropologist
Helsinki, Finland

> *Language is a guide to social reality.*
> —Edward Sapir

The Role of Context in AI's Disruptive Evolution

The rise of generative artificial intelligence (AI) heralds an era of significant technological transformation poised to disrupt various aspects of human civilization. As we navigate this disruption, understanding the concept of context becomes crucial. The inclusion of context when interacting with generative AI is a key technique for effective prompting. However, context is not just an abstract notion but a foundational element that shapes how generative AI systems interact with the world and how we as users interact with these AI systems. Grasping societal context—which includes cultural norms, social behaviors, and ethical considerations—is essential for developing AI that is user-centered, personalized, and as ethically and culturally sensitive as possible. This understanding enables AI to provide the customized and personalized responses that users expect from their AI assistants and collaborators.

For the first time in history, humans face the unique challenge of teaching machines to understand societal context. Unlike traditional programming, which involves coding explicit instructions and relies on explicit rules and instructions, training generative AI involves teaching it to recognize and interpret complex social and cultural cues and, therefore, requires a more sophisticated approach. Understanding what to ask for and how to ask it clearly is crucial for effectively interacting with large language models (LLMs). Understanding the specific area you are inquiring about is essential for formulating relevant questions. Having knowledge in a particular field helps users understand the context of the information they are seeking, which can lead to more effective communication with the LLM. In essence, without domain knowledge, users may struggle to ask the right questions or interpret the responses accurately, limiting the effectiveness of their interactions with LLMs. While technical skills are undoubtedly valuable for understanding model behaviors, such as the impact of training data on responses or token limits, linguistic and literary competencies are also essential for effective prompt engineering. Since AI tools primarily depend on textual input, even slight variations in wording can result in significantly different outcomes.

Design anthropology plays a crucial role here. As a discipline that combines design thinking with anthropological insights, it offers a robust framework for embedding societal context into AI systems. This field focuses on understanding human behavior, cultural practices, and social dynamics, providing a wealth of knowledge that can inform AI design, functionality, and the selection of training data. Leveraging design anthropology ensures that AI solutions resonate with users' social contexts, leading to higher acceptance and satisfaction with the performance of AI.

This chapter emphasizes the crucial role of context and social understanding in the effective use of AI across various industries. As AI systems increasingly influence all aspects of civilization, it is essential for individuals and organizations to recognize the importance of embedding societal, cultural, and contextual awareness into AI applications. This chapter emphasizes the value of design anthropology in helping AI developers, prompt engineers, and users create AI systems that are more accurate, relevant, and culturally sensitive. It suggests that taking societal context into account is a shared responsibility between the AI developers and the end users. By understanding and incorporating the nuances of human context, readers can better navigate AI's disruptive impact, ensuring that AI not only meets technical requirements but also resonates deeply with diverse human experiences, ultimately fostering more effective and ethical AI interactions in their respective fields.

Mastering Prompt Engineering: Foundations and Techniques

Large language models (LLMs), though impressive in their capabilities, possess certain limitations rooted in their reliance on the data they were trained with. These models can generate responses from vast repositories of information, but their knowledge is confined to what was accessible during their training period. This limitation becomes particularly apparent when they attempt to provide up-to-date information or address queries that fall beyond the scope of their training. To overcome these constraints, the technique of prompt engineering has emerged as a critical tool. This method allows developers to guide LLMs toward generating outputs that are more contextually relevant and accurate.

At the heart of prompt engineering are six foundational elements that contribute to crafting an effective prompt: task, context, examples, persona, format, and tone. The task involves clearly defining what is expected from the LLM, typically beginning with an action verb such as "generate" or "analyze." "Context" provides the necessary background information, enabling the model to better understand the situation and the user's specific needs. Including "examples" or templates can significantly improve the quality of the output by offering a clear guide for the model to follow. The "persona" aspect of prompt engineering involves specifying the character or role the LLM should adopt, which can influence both the tone and style of its responses. This is closely related to the format of the output, where the desired structure, whether it be in bullet points, paragraphs, or tables, is indicated. Finally, the "tone" of the response is defined, whether it be formal, casual, enthusiastic, or serious, depending on the nature of the interaction and the intended audience.

Since the effectiveness of these elements depends heavily on the precise use of language, the words chosen to interact with the LLMs must be carefully selected. Linguistics, the scientific study of language and its structure, offers insights into the intricacies of language use. Understanding linguistics is crucial for prompting because it directly influences how language is interpreted and generated in natural language processing (NLP) systems and, thus, enhances the AI's ability to generate relevant and meaningful responses. Various subfields of linguistics contribute to this process, including semantics, which explores the meaning of language, and pragmatics, which examines how context influences language interpretation.

Decoding Context: The Intersection of AI and Design Anthropology

In the realm of AI, context refers to the various factors and conditions that influence how systems interpret inputs and generate outputs. By embedding context, AI systems gain a deeper understanding of the background and specifics of the user's request, which clarifies what the user is seeking. Context enables the AI to tailor its responses to better fit the user's situation, reducing the risk of misinterpretation by supplying the necessary details that inform the AI about the user's needs. This approach ensures that AI systems can comprehend and respond to human inputs in a way that is relevant, appropriate, and sensitive to the surrounding environment

In the broader realm of understanding context, social anthropology—the study of human societies, cultures, and their development—holds significant importance. Traditionally, anthropologists have aimed to interpret the language surrounding a particular subject, analyzing linguistic patterns within their contexts to gain deeper insights into human needs and desires. Post-structuralism examines language within its context to uncover meaning, challenging the traditional view that language has fixed, universally understood meanings. Unlike structuralism, which suggests that word meanings are defined by their relationships within a system, post-structuralism argues that meaning is fluid and context-dependent, reflecting the complexity of language. The meaning of words and phrases shifts depending on context—such as who is speaking, where they are, the surrounding circumstances, and even the historical or cultural background. To truly grasp what someone is conveying, it is essential to look beyond the words themselves and consider the surrounding factors, or "context," that shape what those words mean in that particular situation.

In both AI prompting and social anthropology, context is essential for accurately understanding and interpreting information. It provides the necessary background that informs how a statement, action, or behavior should be understood. In AI, context is often narrowly focused on the immediate interaction between the user and the system, guiding the model to provide the most relevant and accurate response. In design anthropology, context is broader, encompassing cultural, social, economic, and environmental factors that influence how a design will be used, understood, and appreciated by people within a specific community or society.

There is a valuable opportunity to apply anthropological methods within AI prompting. When AI systems demonstrate an understanding of users' cultural and social contexts, users are more likely to trust and engage with them. Additionally, generative AI systems that incorporate such context can continuously learn and improve through user interactions. However, to enable this ongoing learning and improvement, human intervention remains necessary, at least for now. This includes activities such as retraining the model and making algorithm adjustments. By analyzing the context of these interactions, AI systems can adapt their responses over time, becoming more accurate and contextually relevant. This dynamic learning process is essential for maintaining the system's effectiveness in a rapidly changing social landscape.

Integrating Anthropological Insights into AI Development

Incorporating design anthropology into prompt engineering enhances the application of linguistic principles by ensuring that the language used in prompts is not only structurally correct but also culturally and contextually relevant. Semantics, which deals with the literal meaning of words and sentences, ensures that prompts are clear and precise, and convey the intended information accurately. Pragmatics, on the other hand, involves the interpretation of language in context, allowing prompt engineers to design prompts that account for the nuances of meaning, such as implied intentions, cultural references, and the relationship between the speaker and the listener.

Traditional design anthropology methods, such as ethnographic fieldwork, interviews, and cultural probes, enable designers, prompt engineers, and domain experts to gain deep insights into how language is used and interpreted in real-world contexts. For example, participant observation and shadowing during ethnographic fieldwork reveal how different cultures approach concepts like work-life balance. Similarly, diaries, journals, and self-documentation kits allow users to capture their lived experiences, providing valuable data on their natural language use. However, since these methods can be resource-intensive, let's explore more practical approaches to incorporating design anthropological methods specifically in prompting, which requires adaptability and efficiency.

Imagine you are designing a generative AI-powered global customer service chatbot. This chatbot must understand and respond to inquiries about "work-life balance" from users across different cultural contexts. Without proper contextual grounding, the chatbot might provide a generic definition of "work-life balance," simply describing it as the balance between personal life and work, along with generic advice on managing time to balance professional and personal responsibilities.

To effectively apply design anthropological methods in understanding and enhancing the process of prompting a generative AI-powered chatbot, several activities can be considered. These activities are intended to engage developers, end users, and domain experts, ensuring a comprehensive understanding of the chatbot's intended context. One effective approach is organizing participatory design sessions, where the team can observe how users naturally discuss topics such as work-life balance in their own language and expressions. Users might use various metaphors, idioms, or specific terminology that reflect their cultural attitudes toward balancing work and personal life. Understanding these pragmatic nuances can enable the chatbot to respond in a way that resonates with the user's cultural expectations and communication style.

Contextual inquiry is one suggested activity aimed at gathering insights into the real-world scenarios where the chatbot will be utilized. Participants—especially domain experts and end users—could describe their typical work scenarios, focusing on where and how they might interact with the chatbot. This could be accomplished through one-on-one interviews followed by group discussions, with the goal of producing detailed notes on contextual factors, user expectations, and typical use cases.

Another suggestion is persona development, which involves creating detailed user personas that represent the various stakeholders who will interact with the chatbot. Based on insights from the contextual inquiry, the group could collaboratively develop personas that represent different types of end users, such as novice users, expert users, and support staff. The output of this activity would be a set of personas with detailed descriptions, goals, pain points, and expected interactions with the chatbot.

Journey mapping is also recommended as a way to visualize the user journey with the chatbot, highlighting key interaction points. Using the developed personas, journey maps could be created to outline the steps users take to accomplish specific tasks with the chatbot. These maps would note emotions, frustrations, and opportunities for improvement. The result would be journey maps that depict both the current and desired states of interaction with the chatbot, helping to identify critical touchpoints where language and prompts play a crucial role, allowing the team to tailor the chatbot's responses to align with user expectations at each stage.

Additionally, it is suggested that direct collaboration with users be considered to co-create prompts that are culturally relevant and sensitive. This approach ensures that the chatbot's responses are not generic but are specifically tailored to the cultural context of the inquiry. For instance, a prompt developed with input from users in a collectivist culture might emphasize communal support and family considerations in achieving work-life balance, whereas a prompt from an individualist culture might focus on personal time management and self-care. Detailed personas representing various cultural backgrounds and user types can help ensure the chatbot's language is adapted to diverse needs. These personas, developed collaboratively in workshops, could serve as valuable references when designing and testing prompts.

It may also be beneficial to implement adaptive logic in the chatbot that adjusts prompts based on initial user interactions. For example, if a user indicates a preference for a particular cultural context through language choice or self-identification, the chatbot could adjust subsequent prompts accordingly. Context mapping might be used to visualize the various factors influencing user interactions, aiding in the design of adaptive prompts that change based on the user's context, ensuring that the chatbot's responses remain relevant.

The Criticality of Context in the Future of AI

In the rapidly evolving landscape of generative AI, the importance of context cannot be overstated. This chapter has highlighted how integrating design anthropology into the development and application of AI systems can significantly enhance their functionality, relevance, and cultural sensitivity. Context in AI is more than just background information; it is the key to unlocking accurate, meaningful, and user-centered responses.

For prompt engineers and AI developers, the challenge lies in crafting prompts that strike the right balance between providing sufficient context and avoiding information overload. The success of these interactions depends heavily on the careful selection of words and the inclusion of relevant cultural and societal nuances. This iterative process of refining prompts based on AI responses is essential for improving the accuracy and relevance of AI outputs over time. By embedding anthropological methods into the AI development process, we can ensure that AI systems are more aligned with the intricate and diverse realities of the human experience.

About the Author

Growing up in a small town in Finland, Ira Aurora Paavola developed a deep curiosity about the world, one that pushed her beyond the borders of her hometown and into an international adventure at just 14. Moving to Suzhou, China, was more than just a cultural shift—it was an immersion into a tech-driven lifestyle that was years ahead of its time. In China, Ira witnessed how technology seamlessly integrated into daily life, from mobile payments to social interactions, sparking in her a lifelong passion for innovation.

Ira's academic journey led her to study social anthropology in Finland, where she deepened her understanding of human societies. Later, she explored the intersection of design, business, and anthropology during her studies in Hong Kong. Later, Ira's academic path culminated in a second master's degree that focused on human-technology interaction.

Currently, as an AI lead in HR at a multinational company, Ira oversees generative AI adoption, ensuring productivity gains, ethical compliance, and responsible utilization. Ira's mission is to explore how blending design anthropology with AI can shape a future where technology enriches human experiences, fostering a more empathetic, understanding society.

Email: ira.paavola@hotmail.com
LinkedIn: https://www.linkedin.com/in/ira-paavola/

CHAPTER 24

AI MINDSET

By Piero Pierucci, MSC
AI Expert, R&D Leader, Speaker, Author
Rome, Italy

> *We are all agreed that your theory is crazy. The question
> which divides us is whether it is crazy enough to have a
> chance of being correct.*
> —Niels Bohr

To envision the impact of AI on the human mindset, one needs some degree of understanding of what AI is, from the inception of this idea to the current state of affairs. This is a reasonably feasible goal, particularly for those who have been part of this undertaking. Throughout this chapter, I will provide some elements of this journey, demonstrating that the path has not been an easy one. The future of AI can only be predicted with caution, as the margin of error can be quite significant even when viewed from just a year or two in the future.

Another important aspect to observe is the human mindset as it exists today, which is possibly the most difficult point to grasp when trying to understand the impact of AI on it. In fact, there is tremendous discrepancy among experts in defining an apparently obvious concept: intelligence. The assumption that Homo sapiens sapiens, the species to which we all belong, is an intelligent species needs to be justified, better explained, and explored if we want to compare it with the form of intelligence we call artificial.

Biological Intelligence

Let's begin by defining intelligence for a biological entity as the capability to survive in a specific environment. This includes the ability to seek and move towards food, and to escape from dangers. We'll assume that the ability to reproduce is already in place, as it is for simple living entities like bacteria.

If we accept this basic definition of intelligence, we must recognize that even small creatures like nematodes (a primitive form of worm), dating back approximately 600 million years, could sense their surroundings and moving towards or away from beneficial or harmful substances. This was possible thanks to a primitive neural apparatus composed of a few hundred neurons. It's worth noting that the basic structure of a neuron, with its axon and dendrites, has remained essentially unchanged since then (Bennett, 2024).

The next level of biological intelligence emerged around 480 million years ago, when various vertebrates were competing for optimal positions in the evolving marine landscape following the Cambrian explosion. The primitive brain structure of nematodes became much more complex resulting in several naturally occurring variants. The basal ganglia, a structure that evolved to execute and sense movements of the vertebrate's complex body, and the hypothalamus, responsible for the automatic (involuntary) coordination of vital functions like breathing, heart rate, and sleep/wake regulation, found their places. These and many other subsystems are still present throughout all living vertebrates today.

The next significant advancement occurred approximately 200 million years ago with mammals, after some species emerged from the sea. This evolution brought the capability to "imagine" scenarios not yet observed and learn from them. When faced with an alternative, such as choosing between left or right while seeking food, a mammal could construct mental images of both options. It could then decide the best path based on a "feeling" mediated by the amygdala, another part of the brain which further developed in these species to express preferences. This ability to form "images" or "possible scenarios" relied on a primitive form of a new brain region later called the cortex. It's worth noting that this evolutionary leap required enhanced visual capabilities and a warm-blooded system, as the functioning of the new cortex demanded more energy.

Another significant leap occurred with the advent of primates around 15 million years ago. Primates lived in communities of 15 to 50 individuals for improved security against predators. This introduced the necessity for primitive forms of social norms and the need to control individual behavior to secure collaboration.

Within these communities, the ability to form an image of another individual's "intention" and estimate this intention from their behavior began to develop. A more evolved layer of the cortex formed the basis of this capability, known today as the "theory of mind." This allowed primates to build alliances among themselves to gain "power" within the community or to decide whether an individual from a different community was suitable for integration into the group.

The final step is the advent of the Homo genus, with the additional capability of language, which emerged between 2 million and 100,000 years ago. The difficulty in defining a precise timeline reflects the current state of research in this field. However, there is a shared agreement that the appearance of language, and symbolic thinking, placed the Homo genus in a preeminent position compared to other biological entities (Deacon, 1997).

Brain size increased, but language capability seems to be related to the repurposing of brain parts formerly associated exclusively with movement planning. Populations with smaller brain sizes, like Homo Denisovans, still exhibit language capabilities and symbolic communication, indicating that size itself is not the sole factor.

Fire and language, these early technologies, enhanced social capabilities, encouraging group gatherings around the fire. They facilitated the exchange of survival-relevant information, often in the form of storytelling. Additionally, abstract art forms emerged, initially as cave paintings depicting hunting scenes which served to share important practices consolidated in myths and tales. Later, art evolved into decorations on personal objects thus marking distinctions among community members.

Over time, Homo, and later Homo sapiens sapiens, lost certain abilities such as swiftly climbing trees to escape predators, detecting prey through scent, etc. By adopting technology-based solutions, humans compensated for these weaknesses resulting from "unlearning" previously available skills. This new lifestyle ultimately made the species more resilient compared to its ancestors. This shift relied on the species' ability to invent alternative solutions not found in nature itself, but rather obtained through "artificium"—solutions crafted with "art" in the broadest sense.

Each new technology, when deemed useful and widely adopted, introduces cultural changes and, consequently, shifts in mindset. Consider the invention of writing: cultures with left-to-right writing styles tend to favor rational and sequential thinking, a property currently associated with the left hemisphere of the brain. In contrast, right-to-left writing cultures rely more on the associative and holistic capabilities of the right hemisphere (De Kerckhove, 1991).

Automating the Process of Thinking

Returning to AI, we need to recognize it as a collective undertaking, spanning several hundred years of human history. We can trace its roots back to the 19th century, with Babbage's calculating machines and Boole's mathematical theories of logic and probability. At the beginning of the 20th century, Ramón y Cajal shared his theories on the structure of the nervous system, illustrating the behavior of neurons. Around the same period, Pavlov's and Skinner's experimental psychology focused on reinforcement learning mechanisms at work in animals. Subsequent unsuccessful attempts to explain human language based on reinforcement learning mechanisms led to Chomsky's competing theories. These theories, based on the existence of a Universal Grammar, seemed more suitable to explain why children can learn language with relatively few examples: a naturally inherited compositional capability must be in place. Contributions from diverse disciplines—including philosophy, biology, psychology, cognitive science, linguistics, statistics, engineering, and mathematics—formed the critical mass of ideas, concepts, and theories that enabled the simulation of thinking behaviors through machines.

The modern concept of AI, machines exhibiting human-like thinking capabilities, was publicly formulated around the mid-1950s in a series of meetings in the United States, bringing together specialists from various fields. The optimistic idea was that "every aspect of learning, or any feature of intelligence can, in principle, be described so precisely that a machine can simulate it" (Mac Carthy et.al, 1955).

Several outstanding theories were presented and discussed in these meetings:

1. Intelligence can be simulated as a "symbol processing machine" capable of performing deductions based on a number of baseline statements (the logicists).

2. Intelligence can be better simulated as an information processing system, like a black box with inputs and outputs (the statisticians).

3. Intelligence can be better simulated based on a biological model of the brain equipped with evolving capabilities (the neuralists).

The shared purpose was to shed some light on how human intelligence can be simulated, verified, and explained.

Around the 1980s, the great expectations raised by the development of thinking machines were not met by corresponding results. Consequently, funding for research began to decrease, and interest in AI shifted more towards the "weak AI" paradigm—systems capable of performing useful tasks in restricted domains—rather than the "strong AI" paradigm, aimed at designing systems capable of mimicking the human brain, showing competence in a large number of domains and capable of learning from experience. This period became known as the first "AI winter." This shift from the very optimistic vision of the mid-1950s to more practical applications in restricted domains occurred due to repeated failures in achieving the outstanding promises made at the beginning of this adventure. In particular, the idea that intelligence could be reduced to a symbol processing system was recognized as insufficient for practical applications. Formalizing the rules for such a system was a tremendous and possibly unfeasible goal. Systems based on the statistical approach were capable of performing well in certain circumstances, but data scarcity for appropriate training was a limiting factor, and the lack of an explicit model for semantics was a weak point. As far as the neuralist approach is concerned, the community had to wait until processing power became much cheaper to approach a good simulation capability.

At the same time, there was a growing awareness of the complexity involved in brain simulation. The availability of machines capable of simulating "intelligent" capabilities made it possible to verify and support theoretical assumptions from the past, such as unsupervised learning. These machines included:

- Preliminary vision systems

- Speech recognition and synthesis systems

- Systems for natural language understanding in restricted domains

- Robotic systems capable of planning complex tasks in controlled environments

This progress aided the advancement of different disciplines like cognitive science and evolutionary biology, leading to a better understanding of how the human brain:

- Responds to environmental stimuli

- Generates new solutions to novel problems

- Implements efficient processes to simplify the search for complex problem solutions without systematically exploring all possibilities

After this first "AI winter," research focused on "expert systems" capable of performing specific tasks with automation exhibiting sufficient capabilities for real-world purposes. This focus expanded the foundations of AI to include important innovations in:

- Efficient search algorithms

- Logic programming languages and dedicated workstations

- Multi-layered system architectures

- Associated smart tools

These advancements led to applications in various fields, including:

- Medicine

- Automated trading

- Machine translation
- Automated invention
- Face and speech recognition
- Assistance to elder and blind people

The most recent leap in AI is associated with neural networks, a well-known theory from the early 20th century based on a simplified model of the biological neuron. This theory finally found widespread application with the advent of computer hardware, which was initially developed for graphic processing in the gaming industry. Today, we have neural models with sizes on the order of hundreds of billions of parameters, approaching the average of 85 billion neurons in the adult human brain. This advancement has given a tremendous boost to AI at the beginning of the 21st century. Modern applications such as machine translation, computer vision, speech recognition and synthesis, and machines that play chess and Go now demonstrate performance rivaling human capabilities.

The abundance of simulated behaviors and additional tools like fMRI (functional Magnetic Resonance Imaging) have greatly aided the study of the human brain. In parallel, studies in psychology have shown how apparently weak human rational capability can be. When it comes to making important decisions, we often take a "fast thinking path," forged by millions of years of evolution and experiential learning. This is based on challenging situations we may have encountered during our lifespan, allowing us to decide quickly. We apply rational decisions only in circumstances where the energy cost of pursuing a "slow path" (reasoning and focusing solely on that decision while stopping all other activities) is worth the cause. This rational capability is rather lazy, likely due to evolutionary purposes (Kahneman, 2013).

AI Mindset

So where do we stand today? These are a few questions that I hear frequently, along with my best possible answers.

Before questions and answers, an invitation: when we talk about AI, we should not "personify" it. While chatting with a large language model we should not forget that we are actually talking with a big mirror, created by a large community of humans. Think of it as a democratized, simplified, and interactive version of Diderot's encyclopedia, with somewhat less quality, but cheaper, and continuously evolving (Dennis Yi Tienen, 2024). We are not chatting with an alien form of "intelligence." We are talking with us, we can see the good and the evil of humanity, and this can increase our level of awareness.

Will AI Influence Our Mindset?

As already happened with other technologies we invented, the answer is yes. We already tend to delegate our memory capabilities to the "web," asking questions in natural language and often trusting the responses without necessary fact-checking, as this would be too time-consuming. Shortly, there could be cases where we delegate our ability to draft texts for exams or marketing campaigns to machines, assuming we have the final say on what we share as a deliverable product.

This isn't inherently bad, but it's important to be aware of the nature of the system we're using for the draft. We must acknowledge and be conscious of the limitations in the system itself, which is not currently capable of true understanding of the text (or image or translation) being generated. The system may provide incorrect responses (hallucinations) simply because it was trained on a huge dataset composed of both good and bad materials. Controlling the quality of the outcome through human checking (RLHF—reinforcement learning from human feedback) is a costly exercise, currently delegated to low-paid workers in developing countries, and this approach doesn't scale well.

In perspective, it is my opinion that a new basic system architecture is needed; the Transformer (the winning architecture we have today across AI applications) is not enough. In a few years from now—maybe 10, maybe more—a new assessment will take place.

Is AGI (Artificial General Intelligence) around the Corner?

Again, considering the long path that made us "intelligent" as biological systems and the fact that most animals share elements of a similar brain structure, we can say that the structure of the body and the nervous system we have developed is the more efficient thing to survive on earth. Some AI researchers do agree that without embodied intelligence, we face a significant challenge in making the machines understand the world; we have tried hard in the past and failed, but the quest continues in this direction. For those who enjoy pure speculation, I think that should a "superintelligence" emerge based on silicon, it would likely not make favorable decisions for us. It would become an obstacle more than a useful thing; therefore, we would reject it.

Will AI Be Dangerous for Us?

Unfortunately, this may happen, but not because machines will become more intelligent than us and prevail. The danger lies in how we use AI today, including:

- Military applications
- Social control applications
- The growing cost of maintaining huge and energy-hungry data centers
- Effects on the labor market

These factors could become unsustainable for the human community. It's a matter of prioritizing where we allocate resources, and I trust in the human species' common sense to make the right decisions. We need also to help younger generations understand both the benefits and drawbacks of what we call AI.

Can AI Be Helpful to Us?

My answer is a resounding yes. It has already helped tremendously in understanding how our initial hypotheses about the human brain's functioning were incorrect and how complex the matter truly is. We can safely say that today we have a much better understanding of how effective biological intelligence is in interacting multidimensionally with the world around us (through vision, hearing, taste, smell, and touch), taking fast decisions, and acting upon them. What we know as a fact is that AI and diverse scientific disciplines enjoy mutual influence.

With recent discoveries in neuroscience, we are better positioned to understand how non-conscious decisions made by natural intelligence are spanning through the entire spectrum of thinking and action planning (Kahneman, 2013).

Thanks to recent biology research, we are even more aware that intelligence is embodied not only in the control center (the brain) but dispersed throughout the entire body, so that we can make good decisions also based on "gut feelings." We recognize that real neurons are not as simple as described in current neural models running on hardware. This brings us closer to explaining why intuition is based on something more complex than a network of simplified neurons. Real neuron communication can be mediated, in some body regions, chemically rather than electronically (Damasio, 2018).

In conclusion, we can be grateful to AI research for this increased awareness, and we should not forget to closely monitor its applications. We should also recognize that automation has reduced barriers to knowledge acquisition and access, especially when compared to the past, and this is also a positive value we need to enforce.

References

Bennett, 2024: Max Bennett, *A Brief History of Intelligence: Why the Evolution of the Brain Holds the Key to the Future of Ai* (Harper Collins, 2024).

Damasio, 2019: Antonio Damasio, *Strange Order of Things* (Random House, 2019).

Dennis Yi Tienen, 2024: Dennis Yi Tienen, *Literary Theory for Robots: How Computers Learned to Write* (Norton Short, 2024).

De Kerckhove, 1991: Derrick De Kerckhove, *Brainframes: Technology, Mind and Business* (Bosch & Keuning, 1991).

Kahneman, 2013: Daniel Kahneman, *Thinking Fast and Slow* (Farrar Straus & Giroux, 2013.

MacCarthy et.al, 1955: https://www-formal.stanford.edu/jmc/history/dartmouth/dartmouth.html.

About the Author

Piero Pierucci, MSC, is an Italian AI expert with a background in electronic engineering and a strong passion for language. Since 1986 he has been exploring the mysteries of language through technology. He initially served as a research scientist in multinational private R&D labs like IBM and Alcatel, where he pioneered the adoption of HMM (Hidden Markov Models) in the domain of automatic TTS (text to speech) synthesis systems.

Piero served as the program director for speech technology applications at Bontempi, a renowned Italian music industry company, where he integrated the power of speech synthesis into digital musical instruments. He co-founded the electronic art collective Plancton Art Studio to explore human communication at the intersection of science and art. The studio participated in several prestigious art venues, including SIGGRAPH and Imagina.

Piero relocated to Switzerland to join SVOX AG, a startup in Zurich, where he served as head of product development for a compact TTS system based on HMMs. This system was later adopted by Google as FreeTTS in its nascent Android operating system. Following the merger and acquisition of SVOX by Nuance Communications, he served as the technical lead for the core TTS unification project, building the largest TTS system in the world, later on acquired by Microsoft

Piero assisted another Swiss startup, Telepathy Labs GmbH, in building a team from the ground up and developing a digital agent technology platform. This foundation supports virtual agent business applications based on the latest AI deep learning models.

After more than 15 years abroad, Piero has returned to his home country, Italy. He is now contributing to Mediavoice s.r.l. in Rome, developing AI products for the visually impaired and preparing them for the Silicon Valley market.

Email: geosound@mclink.it
LinkedIn: https://it.linkedin.com/in/piero-pierucci

EMBRACING AI: CULTIVATING THE RIGHT MINDSET AND SKILLS

By Marcin Połulich
Operations & Growth, AI Expert and Advisor
Wroclaw, Poland

I am deeply grateful to my parents for their unwavering support and dedication, which enabled me to pursue my education and achieve my goals. Their love and guidance throughout my upbringing have been invaluable, and I am thankful for their many sacrifices. I dedicate this chapter to them, as a testament to their encouragement and belief in me, and I owe my success to their endless love and support.

If somebody offers you an amazing opportunity but you are not sure you can do it, say yes—then learn how to do it later!
—Richard Branson

Artificial intelligence (AI) has been a subject of fascination for both experts and the general public for a long time. It has transformed from a hypothetical idea into a fundamental component of technology. The concept of AI can be traced back to ancient myths and philosophical debates about intelligent machines. Nevertheless, as we know it today, AI started to emerge in the middle of the 20th century.

A key figure in the establishment of AI is Alan Turing—an English mathematician, scientist, and soldier. I consider him the father of the field. In 1950, Turing released his influential paper, "Computing Machinery and Intelligence," where he famously raised a question: "Can machines think?" In this paper, he introduced to the world what is now known as the Turing Test, a check of a machine's ability to demonstrate intelligent behavior equivalent to human beings'.

His concepts provided a foundation for understanding machine intelligence and set the stage for future research in AI.

During World War II, Turing significantly contributed to breaking the Enigma code. The British team, led by him, took the initial work delivered by the Polish mathematicians Marian Rejewski, Jerzy Różycki, and Henryk Zygalski, and expanded on that, developing new techniques to crack German cryptographic systems.

In 1956, the Dartmouth Conference officially introduced the term "artificial intelligence," signaling the start of AI as an academic field. Pioneers such as John McCarthy, Marvin Minsky, and Allen Newell expanded on the groundwork established by Turing. The initial emphasis was on developing machines capable of executing tasks that required human-like intelligence, such as playing chess or solving winding problems. These symbolic AI systems experienced remarkable early achievements, which generated a strong belief that true machine intelligence was within reach. However, the 1970s brought about a period known as the "AI winter," slowing down the progress due to hardware limitations. Despite this setback, researchers, inspired by the pioneering accomplishments of Turing and others, continued their endeavors.

AI experienced a revival in the 1990s and early 2000s, driven by the emergence of machine learning and eventually deep learning. The combination of large datasets and greater computational power allowed AI systems to analyze large volumes of data, resulting in significant progress in domains such as image recognition and natural language processing.

Nowadays, AI is deeply integrated into everyday life, from virtual assistants to autonomous vehicles. As we advance, it is crucial to appreciate Turing's early contributions, which continue to influence the ethical and technical challenges AI faces today. I think his vision of intelligent machines remains fundamental to the progression of AI.

Current State of AI Technology (In 2024)

The rapid advancement of artificial intelligence has introduced a range of technologies that are revolutionizing industries and daily life. From machines that learn from data to those that comprehend human language, AI technologies have become indispensable tools for addressing complex challenges.

This section explores the most impactful AI technologies that are shaping the contemporary world, lying at the heart of AI's continuous evolution and its growing influence across various sectors.

Machine Learning (ML)

At the core of modern AI, machine learning enables systems to learn from data and enhance their performance over time. It powers a wide array of applications, from image recognition to personalized recommendations, forming the foundation for many AI-driven solutions.

Deep Learning (DL)

As a subset of ML, deep learning utilizes artificial neural networks with multiple layers to model intricate patterns. It excels in tasks such as natural language processing, computer vision, and autonomous systems, pushing the boundaries of AI capabilities.

Reinforcement Learning (RL)

RL involves learning through interaction with an environment, optimizing decisions based on rewards or penalties. This approach is commonly applied in robotics, game AI, and autonomous driving, enabling AI systems to adapt and improve their performance in dynamic environments.

Natural Language Processing (NLP)

NLP empowers machines to understand, interpret, and generate human language. From chatbots to voice-activated assistants, NLP has become crucial for enhancing human-computer interaction and bridging the communication gap between humans and machines.

Explainable AI (XAI)

Explainable AI focuses on making AI systems more transparent and understandable to humans. This is particularly important in critical areas like healthcare and finance, where understanding the reasoning behind an AI system's decisions is vital for building trust and ensuring accountability.

Generative AI

Generative AI models are designed to create new data that resembles existing data. These models are utilized in creative industries for art, music, and video generation, and are also applied in areas like drug discovery, expanding the creative and problem-solving capabilities of AI.

Computer Vision (CV)

Computer vision allows AI systems to interpret and understand visual data from the world, such as images and videos. It finds applications in facial recognition, autonomous vehicles, medical imaging, and numerous other fields, enhancing machines' ability to perceive and analyze visual information.

Edge AI

Edge AI involves deploying AI algorithms directly on devices like smartphones, cameras, or internet of things (IoT) sensors, rather than relying on cloud computing. This approach enables faster decision-making and reduces latency, which is crucial for real-time applications such as autonomous drones, smart home systems, and industrial automation.

AIoT (Artificial Intelligence of Things)

AIoT combines artificial intelligence technologies with internet of things (IoT) infrastructure, creating more efficient IoT operations. It enables connected devices and systems to analyze data, make decisions, and act autonomously without human intervention. AIoT is applied in smart homes, industrial settings, and autonomous vehicles, enhancing data management and improving human-machine interactions.

Machine Unlearning

Machine unlearning is an emerging technique that allows for selective removal or modification of specific information from trained machine learning models. This capability is crucial for addressing privacy concerns, complying with data removal requests, and adapting models to changing environments without full retraining. It's particularly relevant in applications dealing with sensitive data or those requiring frequent updates to AI models.

Multimodal AI

Multimodal AI refers to systems that can process and integrate multiple types of input data simultaneously, such as text, images, audio, and video. This approach allows AI models to understand and generate content across different modalities, more closely mimicking human-like perception and reasoning. Multimodal AI is applied in advanced computer vision, natural language processing, and robotics, enabling more contextually aware and versatile AI systems.

Adopting the Right Mindset for Using AI

To successfully integrate AI into daily life and business, it's vital to approach it with the right mindset. While AI offers new possibilities, understanding its strengths, limitations, and ethical implications is essential. This section outlines how to adopt a forward-thinking attitude, embrace continuous learning, and ensure ethical AI use.

Understanding AI Capabilities and Limitations

AI demonstrates remarkable abilities in automating tasks, analyzing vast datasets, and aiding decision-making. It excels in pattern recognition, predictive analytics, natural language processing, and autonomous systems. These technologies enable AI to perform exceptionally in specific tasks like disease diagnosis, fraud detection, and personalized customer experiences.

However, AI is not omniscient or fully autonomous. While outperforming humans in certain areas, AI faces significant limitations:

- *Lack of General Intelligence:* AI systems are typically specialized, lacking the broad reasoning abilities of humans.

- *Data Dependence:* AI relies on high-quality datasets, with inaccurate or biased data that can lead to poor outcomes.

- *Limited Contextual Understanding:* AI struggles with nuance, context, and emotions compared to human comprehension.

So it's crucial to view AI as a powerful tool to augment human abilities, not a universal solution. To maximize AI's utility:

- Define clear goals for AI implementations.

- Maintain human oversight, especially in complex decision-making processes.

- Continuously monitor and update AI systems to ensure relevance and accuracy.

Understanding AI's strengths and limitations allows for more effective utilization across various applications and industries. By aligning expectations with AI's actual capabilities, organizations can harness its potential while avoiding unrealistic assumptions about its role and impact.

Embracing Continuous Learning

AI technology is rapidly evolving, with constant breakthroughs transforming innovations into standard practices. As AI becomes increasingly pervasive, staying informed about the latest developments is essential for individuals and organizations. Continuous learning is a key for harnessing AI's potential, as advancements drive competitive advantages. Without staying updated, organizations risk missing opportunities to leverage AI for innovation.

Ethical Considerations in AI Use

The growing integration of AI into society amplifies the importance of ethical considerations. AI's capacity to influence decisions, mold behaviors, and significantly impact lives raises crucial ethical questions regarding fairness, accountability, and transparency. Responsible AI deployment necessitates addressing the potential ethical challenges inherent in this area. As a result, the following aspects have to be taken into account:

- *Bias:* AI can inherit and perpetuate biases from training data, potentially leading to unfair outcomes. Identifying and mitigating bias is crucial for fairness.

- *Privacy:* AI systems processing personal data must adhere to strict privacy standards to protect user information.

- *Accountability:* clarifying responsibility for AI mistakes and ensuring human oversight in critical areas is vital for trust.

- *Transparency:* making AI systems more explainable helps build trust and enables users to understand decision-making processes.

Educating and Empowering Skills for AI

Effective AI adoption hinges on cultivating a diverse skill set that enables individuals and organizations to navigate the dynamic AI landscape. The required competencies encompass both technical expertise and soft skills, mirroring AI's multifaceted nature. By equipping people with these capabilities, we empower them to harness AI's potential and spur innovation across various domains.

Key Skills Needed for AI Adoption

Adopting AI involves a combination of technical expertise and critical thinking, making it crucial to equip teams with a balanced set of skills.

Technical Skills

- *Programming:* proficiency in languages like Python and R is crucial for AI development. This skill enables customization and optimization of AI tools.

- *Data Analysis:* the ability to clean, prepare, and visualize data is essential for identifying patterns that AI models use for predictions or decisions.

- *Machine Learning:* understanding various learning algorithms and frameworks like TensorFlow and PyTorch or OpenCV is necessary for designing and training advanced AI systems capable of processing complex data.

Soft Skills

- *Critical Thinking:* evaluate AI outputs to ensure alignment with business goals. Assess recommendations and identify potential biases or limitations.

- *Problem-Solving:* apply AI solutions innovatively to address complex challenges across various industries. Identify suitable AI approaches for real-world problems.

- *Adaptability:* remain flexible in the rapidly evolving AI field. Embrace new tools and techniques to effectively leverage AI in dynamic environments.

Prompt Engineering

The advancement of large language models (LLMs) like GPT (provided by OpenAI) or Llama (Large Language Model Meta AI) has elevated prompt engineering to a critical skill. This technique involves crafting precise, structured inputs to guide AI towards accurate and relevant outputs. LLMs find applications in various domains, from content creation to automated customer support.

Prompt engineering involves crafting input prompts to elicit precise responses from AI models. The quality of the prompt directly influences the output's relevance and accuracy.

For instance, "Explain AI" generates a broad answer, whereas "Explain reinforcement learning in autonomous driving" produces a targeted response. So effective prompt engineering requires providing sufficient context and direction to guide the AI towards the desired output.

Key Techniques in Prompt Engineering

- *Precision:* craft detailed prompts with specific constraints to guide AI responses. For example, request "a 100-word summary" for targeted output.

- *Refinement:* improve prompts iteratively based on initial AI responses. Analyze and adjust, so you will enhance accuracy and relevance.

- *Example-Based Guidance:* include sample outputs in prompts to steer AI towards desired formats or styles.

- *Contextual Framing:* embed relevant context in prompts for more focused answers. For instance, ask about "AI benefits in healthcare" rather than general AI benefits.

- These strategies will optimize AI interactions, allowing for more accurate and useful results.

AI Case Studies from Various Industries

AI has demonstrated its ability to transform a variety of industries by streamlining processes, improving decision-making, and delivering more accurate results.

In this section, we explore three real-world case studies from various industries, illustrating the impact of AI-driven solutions.

Case Study 1: AI in Healthcare—Identifying Inefficiencies in Patient Diagnosis

Healthcare systems often face challenges with diagnostic accuracy and speed, leading to delayed treatments and higher costs. The traditional diagnostic process, heavily reliant on manual review of patient data and medical images, can be slow and prone to human error, resulting in missed diagnoses or delayed care.

To address these inefficiencies, AI-driven diagnostic tools were introduced to assist doctors in analyzing medical images and patient data. Machine learning algorithms were trained on vast datasets of medical scans, enabling them to detect patterns indicative of diseases like cancer, heart disease, and neurological disorders. These tools are capable of processing data much faster and accurately than human doctors and can flag potential issues for further review.

The implementation of AI diagnostic tools led to significant improvements in diagnostic accuracy, especially in detecting early-stage diseases that may have otherwise gone unnoticed. Additionally, AI systems helped reduce the time needed to diagnose patients, limit diagnostic errors, and improve overall healthcare efficiency.

Case Study 2: AI in Finance—Enhancing Fraud Detection

Conventional fraud detection methods, which typically depend on preset rules and manual oversight, often fall short in identifying novel and evolving fraudulent strategies. As perpetrators become increasingly adept, these traditional systems struggle to maintain effectiveness, potentially leading to financial losses and reputational harm for organizations.

To enhance fraud detection capabilities, AI-driven algorithms were deployed to scrutinize transaction patterns and flag suspicious activities in real time. These advanced machine learning models, trained on extensive historical transaction datasets, were utilized to identify atypical patterns potentially indicative of fraud. Such patterns might include a series of small transactions occurring rapidly or transactions originating from unexpected locations.

AI-driven fraud detection significantly reduced fraudulent incidents. Unlike traditional methods, AI models adapt to new fraud tactics, leading to fewer false positives and improved security for financial institutions. These systems enabled earlier fraud detection while reducing resources needed for manual monitoring and increased success-ratio.

Case Study 3: AI in Manufacturing—Optimizing Production Processes
Manufacturing sectors encounter difficulties in streamlining production lines to reduce downtime and boost efficiency. Unexpected equipment breakdowns and suboptimal processes can lead to substantial delays, affecting output and escalating costs. Anticipating machinery failures and identifying necessary production line adjustments are crucial aspects of maintaining operational effectiveness.

AI-powered predictive maintenance solutions were implemented to enhance production efficiency. These systems utilize sensors and IoT devices to continuously monitor machinery, analyzing data such as temperature, vibration, and pressure to anticipate potential equipment failures. Furthermore, AI-driven optimization models were employed on production lines to pinpoint inefficiencies and suggest improvements for streamlining processes.

Leveraging AI for predictive maintenance enabled manufacturers to significantly reduce unscheduled downtime by proactively identifying and addressing potential issues. This strategy minimized disruptions and extended machinery lifespan. AI-powered process optimization also enhanced production efficiency, increasing output and reducing costs. AI incorporation boosted productivity and fostered a more resilient and adaptable manufacturing process.

Challenges and Future Directions

Implementing AI in organizations comes with several significant challenges. One of the biggest issues is data quality and availability. AI systems require large amounts of high-quality data to function properly, but many organizations struggle with data that is incomplete, biased, or difficult to access. Ensuring that data is accurate, representative, and readily available requires strong data governance practices.

Another challenge is the integration of AI with existing systems. Incorporating AI into an organization's current infrastructure and workflows can be complex. It involves making sure that AI systems are compatible with legacy IT systems and ensuring smooth data flow between AI and non-AI components. Additionally, business processes often need to be adapted to fully leverage AI's potential, which adds another layer of complexity.

Workforce adaptation and resistance is also a major challenge. The introduction of AI frequently leads to changes in job roles and skill requirements, which can cause resistance among employees who may fear job displacement or feel unprepared to work with AI technologies. Addressing these concerns through effective change management, clear communication, and reskilling initiatives is crucial to fostering an AI-ready workforce.

Final Worlds

The adoption of AI requires a thoughtful approach that balances technology with human capabilities. What I personally believe—success lies in fostering the right mindset, one that views AI as a tool for enhancing, rather than replacing, human skills. This is the AI Mindset.

Organizations must invest in building both technical and soft skills among their workforce, ensuring that employees can effectively leverage AI systems. In this scenario, leadership plays a crucial role in driving AI adoption, creating a culture of innovation, and promoting ethical AI practices.

As I conclude this journey into the world of artificial intelligence, I am reminded of a profound moment during a talk by Mo Gawdat, the former chief business officer at Google X. He stated with conviction, "The genie is out of the bottle." This statement resonates deeply because it highlights the undeniable reality we now face: AI is no longer a distant future but an integral part of our present. The responsibility for how we manage and guide this technology does not lie in the hands of a select few but rests with all of us—as individuals, as professionals, and as a society.

I often return to the iconic imagery of James Cameron's *Terminator* films and the vision of SkyNet. It's a stark portrayal of AI gone awry, one that raises uncomfortable but crucial questions about our future. If we were to ask AI, "What is the greatest threat to planet Earth?"—and if the answer were humankind—how would the machine respond? What actions would follow?

These are not abstract, science fiction musings; they are questions we must grapple with now as we cultivate the right mindset and skills to shape AI's trajectory. The choices we make today will determine whether AI becomes a tool for collective advancement or a force that could potentially threaten our existence. It is our responsibility to ensure that AI reflects our best qualities—wisdom, empathy, and foresight—rather than our darkest fears.

The future of AI is, in many ways, the future of humanity. Let's ensure it's one worth creating.

About the Author

Marcin Połulich is a seasoned leader with many years of experience at the forefront of the tech industry. His expertise spans multiple domains. Throughout his career, he has pioneered advancements in software development, deep tech, and digital transformation, while consistently driving operational excellence and accelerating business growth. As a recognized expert in the industry, Marcin has a proven track record of aligning technology with business objectives, helping companies navigate the challenges of the digital age.

Outside the realm of technology and leadership, he is deeply passionate about sports, regularly challenging himself through physical activities that push his limits and contribute to both his physical and mental well-being.

Email: marcin.polulich@gmail.com
LinkedIn: https://www.linkedin.com/in/marcinpolulich/

CHAPTER 26

EMBRACING TECHNOLOGICAL CURIOSITY

By Christina Rehmeier
Assistant Professor teaching AI, University College Lillebaelt
Hedensted, Jutland, Denmark

If a machine, a Terminator, can learn the value of human life,
maybe we can, too.
—Sarah Connor (Terminator 2: Judgment Day)

I want to start this chapter by saying thank you, dear Hollywood. We got all doomsday prophecies, the story about large computers that can answer for us every question in a pleasant female voice, manipulative robots doing what they were programmed to do in the first place, and a general huge concern for AI.

Fortunately, some portrayals like Jarvis from *Iron Man*, C-3PO from *Star Wars*, and Pixar's Wall-E offer more optimistic views that encourage us to trust AI again.

If we look away from the Western world and turn our focus towards other parts of the world, like Asia, we can see a few things differently. Japan, especially, is known for its use of AI, particularly in a more helpful form, with self-driving cars, robotic arms that can flip pancakes in schools or collecting a car from parts, or even a friendly robotic receptionist who delivers a warm welcome when you enter a building. The thing is that these stories have one thing in common: the AI are all shown as robots.

So, what does all this mean for how we really understand AI? As we dig deeper into this topic, let's not just focus on the robots we see in movies and media. We also need to look at the countless non-robotic forms of AI that are quietly shaping our lives—often without us even being aware of their influence.

The lines that are shaping our lives are beginning to become blurred. With the entrance of certain technology such as ChatGPT and other large language models, and AI's ability to bring conversation to the table that sounds increasingly human, the social implications are what is scaring a lot of people. It has become more difficult to notice these lines and we are now forced to wonder, "Am I chatting with a human or a bot?"

This is where we as human beings need to develop and use our new mindset, the AI Mindset.

A Way of Thinking

So, what sets the AI Mindset apart from our traditional ways of thinking? When we examine the common traditional mindset regarding technology, we often find ourselves accepting information at face value, particularly when it's presented by an authoritative source. This inclination can lead us to overlook the deeper biases in algorithms, limitations, or inaccuracies that AI might be amplifying, like where the AI got its research or literature from.

By embracing an AI Mindset, we're encouraged to dive deeper and think critically. It's not just about what we see, but also we must consider the context and the impact of the information around us. This shift helps us transition from being casual users to more engaged developers of new technologies, like AI. After all, everyday users will play a key role in shaping the future of AI and enhancing its capabilities.

The AI Mindset is all about developing a savvy sense of discernment that helps us navigate the intricate world of AI-generated content. By keeping our eyes wide open, we can confidently find our way through this hazy landscape, ready to spot the subtleties and make sense of the information we experience.

As discussed in Ştefan Trăuşan-Matus's 2024 article "Ethics in AI," we should all think about virtual assistants like Siri or Google Assistant. They are like having a friendly companion right in your pocket, always ready to help out with answers or reminders. Need to set a timer for your cookies or get directions to a new coffee shop? No problem! These AI pals have got your back. But as we revel in this convenience, it's crucial to remain discerning. Just like that well-meaning friend who occasionally mixes up your favorite pizza toppings, these AI companions don't always get it right.

Sometimes, they might misunderstand your requests or deliver information that's just plain inaccurate. And when it comes to IT security, their well-intentioned assistance can also lead us astray. For instance, when we ask our virtual assistant to store sensitive information or manage our calendar, we need to ensure we're not inadvertently putting ourselves at risk. After all, while they work tirelessly to make our lives easier, it's up to us to make sure we're keeping our data secure and our digital lives safe.

Look into our smart home devices. Those clever gadgets that let you control your lights, thermostat, and even your coffee maker with just a voice command or a quick tap on your phone. It's like having a little helper in your home, ready to bring a cozy vibe with dimmed lights or brew your coffee just the way you like it! While it's amazing how these devices simplify our lives and make us feel like we're living in a sci-fi movie, we should also pause for a moment to consider what's happening behind the scenes.

These gadgets constantly collect data about our habits, preferences, and routines, all to serve us better. But remember, they are only as smart as the information they're fed.

A healthy dose of awareness about this data collection will go a long way in helping us navigate this tech-savvy world with both ease and insight. It's like being in a good relationship—communication and trust are key! So, as you enjoy the comforts of your automated home, keep an eye on what it's learning about you.

AI and Human Connection

In our tech-driven era, the rise of conversational AIs is creating a fascinating dynamic in how we connect with one another. These increasingly sophisticated AI systems, like chatbots, virtual assistants, and even AI companions, can engage in conversations that feel surprisingly human-like. This evolution invites us to reflect on the implications and how they can enrich human interactions, but they also have the potential to foster feelings of alienation.

Imagine chatting with a virtual assistant that understands your preferences and responds not just with information, but with empathy. Sounds great, right? These interactions can add a layer of support to our daily lives, offering the comfort of a quick check-in or the ease of automation. However, as we indulge in these digital exchanges, it's essential to consider that they can't fully replace the warmth, depth, and spontaneity of real human connection. We might find ourselves relying on these AIs for companionship without realizing they can't replicate the emotional richness that comes with genuine human interactions.

When we talk about trust in AI, it's a bit of a double-edged sword, according to Fazley Rafys 2024 article "Artificial Intelligence in Cyber Security." On one hand, we trust these systems because they're built on vast amounts of data, often presented by credible sources. The confidence we place in them comes from their ability to provide answers or suggestions that feel informed and timely. Yet, this trust can be easily eroded by instances of AI failure, whether it's misunderstanding our requests or making recommendations that miss the mark. This is where our own mindset comes into play, encouraging us to engage critically and maintain a level of discernment about when and how we place our trust in these digital entities.

Humans Embracing AI or AI Embracing Humans?

Rather than perceiving AI as a mere substitute for the human touch, we can consider it an invaluable ally. This concept of augmented intelligence underscores the potential for AI to enhance human capabilities instead of diminishing them. In fields like medicine, for instance, AI systems are making significant contributions by supporting healthcare professionals in diagnosing diseases with remarkable accuracy. By analyzing extensive datasets and identifying patterns that may elude human observation, AI empowers doctors to deliver improved patient care. This partnership exemplifies how human ingenuity, when combined with intelligent technology, can unlock remarkable advancements—while also emphasizing the importance of IT security measures to protect sensitive patient information and maintain trust in AI systems.

As we navigate an increasingly tech-driven world, it becomes evident that AI literacy should serve as a fundamental component of our education, workplaces, and communities. Understanding AI equips individuals to maneuver through this evolving landscape with both curiosity and confidence.

Schools can introduce AI concepts early on, teaching students not only the mechanics but also the underlying principles and ethical considerations of technology, including the implications for data privacy and security. Workplaces can offer training programs that empower employees to leverage AI tools while also addressing best practices for safeguarding sensitive information, thereby highlighting how these technologies can enhance productivity and job satisfaction rather than rendering human roles obsolete.

Moreover, community programs can foster open dialogues about AI, incorporating diverse perspectives to explore its potential benefits, implications, and challenges. This inclusive approach can facilitate the transition from casual users to proficient masters of AI, with an awareness of security protocols that protect both users and data integrity.

By embracing an AI Mindset, we cultivate a culture that values inquiry and responsible engagement with technology, underscoring the necessity of implementing robust IT security practices. With both awareness and enthusiasm, we can harness the potential of AI while safeguarding the irreplaceable essence of human connection and the security of our digital environments.

In this exciting partnership between humans and AI, there's plenty of room for both to thrive. Together, we can create a future where technology embraces our connections rather than takes away from them, all while keeping security and ethical values at the forefront.

Be Curious but Skeptical

As we dive into the exciting realm of AI and technology, it's essential to blend curiosity with a dash of skepticism. Curiosity fuels innovation and opens our minds to new possibilities while skepticism helps us keep our wits about us, ensuring we critically consider technology's impact on our lives and society. Balancing these two traits allows us to engage with technology in a fun, thoughtful way.

When exploring AI, it's important to go beyond the "how" and delve into the "why" and "what if." Take a moment to wonder whether that shiny new app truly keeps your data safe or if there are hidden biases lurking in its algorithms. Often, we struggle to accurately assess the true benefits of a technology and where it can be effectively used. For example, consider facial recognition technology. While it promises enhanced security and convenience, its use in public spaces raises serious concerns about privacy, potential misuse, and biases against certain demographic groups. Recognizing such nuances can help ensure that technology serves us rather than the other way around.

Diving into the world of AI can be both exciting and rewarding! Consider enrolling in online courses that cover AI fundamentals as well as the ethical implications of technology. This knowledge won't just satisfy your curiosity; it will also equip you with the tools to navigate the tech landscape with a discerning eye, helping you make better judgments about where and how these technologies can be beneficial.

Engaging in discussions about technology and ethics, whether online or in your local community, can further enhance your understanding. Sharing perspectives with others who are passionate about responsible tech often uncovers new ideas and insights.

Don't shy away from experimenting with AI tools in your daily life. Whether you're using them for creative projects or productivity boosts, tinkering with technology can help demystify it and help in the process to move from being casual users to more engaged developers of new technologies. Just keep those crucial privacy settings in mind and reflect on how your data is being handled. It's all part of being a savvy user in a world where our assessments of technology's benefits can sometimes lead us astray.

As you journey into this exciting tech future, keep that balance of curiosity and skepticism in mind, for these are critical elements in the AI Mindset. Embrace your role as a knowledgeable explorer while also advocating for ethical practices, data security, and the responsible use of technology. The future is bright, and it's up to us to ensure it shines the right way.

About the Author

Christina Rehmeier is a creative generalist and skilled educator with a passion for technology and communication. With eight years of teaching experience and five years in welfare technology and IT projects, she brings a unique blend of creativity and technical expertise to her work. Christina's background in web development and IT security informs her innovative approach to education and project management.

In recent years, Christina has embraced artificial intelligence, teaching the subject at a high level and focusing on machine learning fundamentals. Her ability to explain complex concepts in accessible ways has made her a valued instructor in this emerging field.

A mother of two and devoted wife to Mark, Christina believes in wholehearted living and continuous learning. Her curiosity and commitment to personal growth drive her to stay at the forefront of technological advancements, inspiring others to embrace innovation and lifelong learning.

LinkedIn: www.linkedin.com/in/christina-rehmeier-75597b26

GEN-AI AND EMOTIONAL INTELLIGENCE FOR IMPROVED WORKFORCE PRODUCTIVITY

By Kathryn Simons-Porter
Customer Exp. & Emotional Intelligence, AI Consultant
Leeds, England, United Kingdom

*Written in dedication and love for my late mum, Judith.
Always to be found with her nose in a book and
eager to learn.*

Prelude—The Case of the Blunt Blade

Mark, a brilliant engineer, excelled with numbers and logic, his mind always whirring with calculations and problem-solving. But Mark struggled with people. His team admired his technical skills yet dreaded his harsh critiques. For instance; Tim's innovative idea was dismissed, and Bill's tireless efforts were met with condescension. The team was veiled in fear of "doing the wrong thing".

After only six months in his post, Mark's team's morale plummeted, impacting productivity and quality. During a crucial client presentation, Mark's technical focus and lack of empathy caused the client to walk out. The once vibrant atmosphere was now filled with tension and fear of failure. Productivity was taking a hit, and the quality of work began to suffer.

The team realised Mark's emotional shortcomings were damaging not only their spirits but also the company's reputation, and sales were at risk—whenever Mark was involved.

Change was imperative. Whether Mark could develop emotional intelligence or the company would need to adapt remained uncertain. The future of both Mark and the team hung in the balance.

GenAI and Emotional Intelligence (EI)—Improving Workplace Productivity

Research-based training products and the service company TalentSmartEQ state there are inherent links between high performers and high levels of EI across many roles within organisations—sales, engineering, law enforcement, HR, leadership. According to TalentSmartEQ, 90% of the top performers that they have worked with are also high in emotional intelligence.

But here's the thing—now that the planet has been gifted with the power of generative AI, can the cultivation of people at work be further propelled through innovation?

This chapter looks at how GenAI used in the workplace can cultivate and develop greater levels of emotional intelligence, leading to more effective ways of working, greater levels of productivity, and a faster delivery of business outcomes. We will examine how individuals (in the case of the Blunt Blade story—how "Mark") can use GenAI EI innovation to help them understand their emotions, and more effectively communicate with their team (and their customers).

First, let's consider—what do we mean by emotional intelligence? The term "emotional intelligence" (otherwise known as "emotional quotient" or EQ) was first coined in 1990 by psychologists Salovey and Mayer. However, the concept of EI gathered attention in 1995 with the publication of Daniel Goleman's book *Emotional Intelligence: Why it can Matter More Than IQ*. Essentially, EI is the ability to recognise, understand, and manage your own emotions whilst effectively navigating social interactions by recognising and influencing the emotions of others.

Organisations that value the importance of EI in the workforce have cited high benefits as early as 1990, as documented in Harvard Business Review's "What Makes a Leader". More recently companies such as AWS have invested in EI training with their EPIC leadership program (EPIC stands for "empathy," "purpose," "inspiration," and "connection") and upskilled 350,000 Amazonians worldwide. Cisco's Splunk was inspired by their work and adopted the EPIC program to enable their leaders in EI skills, and are currently rolling this out to hundreds of employees.

Now let's turn to AI. AI automation has existed in the workplace for some years, removing mundane tasks, improving operational efficiencies and lowering cost. Examples include robots used on car production lines and automation used in the back office to process insurance claims. Traditional AI performs tasks, following rules laid down by a human. According to Gartner Management Consulting (www.gartner.com), by 2026 80% of organisations will have implemented GenAI for the purpose of improvements to operational efficiency.

However, then came along GenAI. In 2019, Open AI launched the first generative AI (GenAI) long language model (LLM), and, arguably, a revolution was launched. Unlike traditional AI, GenAI creates new content like text, images, or music, and gives guidance and learning—often based on patterns learned from existing data.

In a contact centre scenario, there are now some spectacular agent tools that use GenAI to feed the agent with EI-type cues. For instance, it might instruct the agent to "demonstrate ownership." The intended goal of a contact centre is to improve customer experience, whilst reducing operational cost; therefore, when the GenAI feeds the contact centre agent EI-"type" instructions, the aim is that customers will have better experiences.

However, there are two clear disadvantages of using GenAI to instruct agents:

1. Those contact centre agents already with high EI find the instructions to be unnecessary micro-management, and they disengage, which results in a decline in productivity.

2. By forcing (rather than coaching) the agent to say a particular phrase, the agent's Disingenuity is apparent to the customer, which is potentially counterproductive.

So, is using GenAI to boost or develop EI a paradox? Can apparently emotionless AI assist in enhancing EI?

Let's go back to Daniel Goleman (arguably the grandfather of EI). Goleman believed that successful leaders possessed higher levels of EI. Goleman stated that there are four cornerstones of EI: self- awareness, social-awareness, self-regulation—all of which lead to the end goal of—effective relationship management.

Now, let's bring together GenAI and also EI. We'll start by defining Goleman's four cornerstones of EI and then discuss GenAI's relationship to each.

1. *Self-Awareness*: we are aware of our emotions, know how we're feeling, and can clearly express ourselves.

2. *Self-Management*: we can use our self-awareness to take appropriate action and manage our emotional state (often called self-regulation).

3. *Social Awareness*: we use our empathy to understand how other people are feeling and respond to their emotional state.

4. *Relationship Management*: we can take our emotions and the emotions of others and positively manage and influence social interactions.

Self-Awareness

"Mirror mirror on the wall …" Intuitive GenAI can become that mirror. GenAI can provide multi-channel analysis across email, voice, and digital to provide feedback on communication styles, effectiveness, and tone of voice. Taking that one step further, wearable devices can monitor body language and facial expressions, all of which provides insights into emotional states and measures an individual's reactions. Why is this important? Self-awareness leads to more efficient collaboration across teams, greater levels of relationship building, and effective leadership—all of which "Mark," the engineer in our story at the beginning of this chapter, lacked. Mark's lack of self-awareness caused his team to feel intimidated, morale and motivation crashed, and essentially Mark was just disliked—all of which are detrimental to workforce productivity.

GenAI personalised development plans are a tool that can identify an individual's strengths and weaknesses, and highlight development areas based on performance data, allowing for tailored self-improvement plans. By analysing aspirations and career paths, GenAI can suggest specific goals to enhance self-awareness and personal growth.

Self-Management

GenAI EI can be used in a variety of ways to help with self-management.

Stress and Anxiety Reduction: GenAI EI can provide real-time interventions such as mindfulness exercises, breathing techniques, or calming music to help a person manage their stress and anxiety levels.

Mood Enhancement: by suggesting activities or content that align with an individual's preferences and emotional state, GenAI EI can help elevate mood and promote well-being.

Goal Setting and Tracking: GenAI EI can assist someone in setting SMART goals (specific, measurable, achievable, relevant, and time-bound) and provide personalised reminders, encouragement, and feedback to keep them motivated.

Personalised Rewards and Incentives: by understanding individual preferences, GenAI EI can suggest rewards or incentives that are truly motivating, fostering a sense of accomplishment and progress.

Relationship Management

This is especially important for leaders within organisations as they grapple with the new AI era. Certainly, at the time of this writing, GenAI might be on the "to-do list" of many leaders, but how, where, and exactly why they need to use GenAI remains a mystery to most. By utilising GenAI EI coaching models, leaders can become more self-aware and more trusting of their AI-enabled, autonomous direct reports.

Furthermore, GenAI EI can help individuals with decision-making. Emotionally intelligent leaders are better equipped to make sound decisions under pressure, considering emotional and logical factors.

In relation to conflict resolution, GenAI EI can offer suggestions on how to approach and resolve conflicts effectively, taking into account the emotional dynamics involved.

Social (Organisation) Awareness

GenAI EI models can analyse and interpret the emotional tone of communications, such as email and chats, meeting notes, etc., helping employees understand each other better and reducing misunderstandings.

These models can suggest alternative phrasing or offer insights on how to improve communication based on the emotional context. Furthermore, they can also facilitate brainstorming and ideation sessions by generating creative ideas and solutions based on the emotional needs and preferences of the team.

Neurodiversity in the Workplace

In recent years, it would appear that neurodiverse people (individuals whose brains function, learn, and process information in ways that differ from what is considered "neurotypical" or "typical") are being more welcomed at work. As a stepmom to an autistic young man currently at university, it is reassuring that many organisations have now incorporated total diversity into their people strategy and that my stepson will have more easily accessible career opportunities and the ability to work in an environment that offers the right infrastructure to help him flourish within this role and enable him to give his best to his employer.

However, there is still some way to go. Both autism spectrum disorder (ASD) and attention-deficit/hyperactivity disorder (ADHD) are classified as mental disorders in the *Diagnostic and Statistical Manual of Mental Disorders* (DSM-5), the widely used diagnostic tool published by the American Psychiatric Association (APA). The DSM contained (and still does) quite a long list of "mental disorders." It was only in 1973 that the APA had homosexuality removed from the list.

Another consideration when creating people strategy is around pathological demand avoidance (PDA). While not yet officially recognised in diagnostic manuals like the DSM-5, PDA is increasingly understood as a profile within the autism spectrum. It's characterised by an intense and pervasive resistance to everyday demands and expectations, even those the individual might want to fulfil. So, going back to the example of agents in a contact centre, literally instructing agents to say or do a task might have the opposite effect—they refuse to do it, or do it well. Therefore, guidance and coaching is surely the answer, as opposed to instructing and mandating from a script.

As already discussed, tailored feedback generated from GenAI can offer a personalised critique that considers individual strengths and challenges, helping neurodivergent employees understand their performance areas. GenAI EI can also be used to create virtual assistants that offer emotional support and guidance, helping a person to navigate their working environment. A GenAI virtual assistant can become a personal coach that never goes on holiday or has other clients to see, one that is there to help you when you can't sleep at 4am due to overthinking a presentation the following day. Such models can be trained to understand and adapt to different communication styles, which can aid in internal collaboration and decrease misunderstandings.

Creating a workplace where neurodivergent employees can work with self-confidence also creates a more blended and diverse working community, which in turn creates greater levels of productivity and engagement. And when such development tools are offered to all employees, it helps to promote a foundation at work where everyone is welcome and equal.

GenAI EI for Job Screening

A truly productive workplace should be one that holds responsibility and purpose at its core. GenAI EI can aid with this by offering continuous and ongoing development and improvement to ensure that individuals indeed flourish together and alone in their work.

While someone with very low EI can be developed over time, individuals and their leaders that flourish probably had high EI at the time of hiring. A great amount of time, effort, and cost can be saved by hiring the right people, people that fit the bill for the hard skills—but also that demonstrate high EI. GenAI offers possibilities for identifying EI during hiring, at the first screening (for instance). Please note that caution must be applied from an ethics and bias consideration. A combination of AI analysis (e.g., using AI in video interviews to spot certain non-verbal cues, body language, etc.), simulating interactions (role plays with "customers" or an assessment on how they behave in a team), and other soft skills assessment can provide a comprehensive understanding of an individual's EI level.

Ethics and Bias

Bias in generative AI models refers to the presence of unfair or stereo-typical generated content. This bias usually results from the training data and reinforces prejudices or societal inequalities present in the data. GenAI often falls under criticism due to potential biased outputs, ethical concerns relating to fake or misleading information, and the need for greater transparency in the development of AI-generated content. Because of this, if an organisation is going to invest in GenAI EI in order to cultivate their people, they need to make sure that the content and guidance is genuine, accurate, correct, and unbiased. Organisations need to be able to put trust in the investment, much like they would with any training platform or resource—human or otherwise—to ensure a more fair and equitable workplace for everyone.

It should be noted that at the time of this writing, GenAI EI models are still limited by the data they are trained on. If the training data is biased, the model may still produce biased outputs. Furthermore, bias can be subtle and multifaceted, making it difficult for even the most sophisticated GenAI EI models to detect and address all forms of bias. Human-in-the-loop is, therefore, necessary to ensure that GenAI EI models are used ethically and responsibly, and to address any biases that might arise.

Adoption by the Workforce and Overcoming Fear

To some, the opportunity to hold up a mirror and observe oneself might seem curious—liberating even. However, to many, this is scary, intimidating, and humiliating. As ever, technology should be treated as a tool, in this case for the development of self and the harmony of teams. Younger generation workers (e.g., Gen X and Millennials) embrace GenAI more readily (typically) than Boomers because they have grown up with digital technology. They also might have entered into the workplace using AI in their everyday jobs (e.g., AI workflows), and they are typically more open to change. In any case, leaders and HR teams need to "sell" the idea to their colleagues that GenAI EI is being used to develop and improve—to fine tune—their current skills, and not to fix flaws. A positive adaptive mindset is crucial here to avoid the "robots are taking our jobs" perspectives.

Organisations also need to consider that to cultivate the best from people perhaps they should set benchmark standards at average performance and output. Even with the best GenAI EI intention and investment, no workforce is going to perform to their optimum every single day. These North Star goals are unrealistic and will have the reverse impact, i.e., disengagement, low motivation, burnout, absenteeism, high attrition, and all the costs associated.

A Turning Point

Did you wonder what happened to Mark and his long-suffering team?

The disastrous client meeting cemented Mark's reputation as a brilliant but abrasive engineer. The company faced a PR crisis, morale further plummeted, and Mark, oblivious, proposed a technical report as a solution. The CEO, recognising Mark's value, implemented mandatory EI training, for all employees.

Mark, initially resistant, found himself intrigued. He began applying his analytical skills to understanding human behaviour, becoming more self-aware. Slowly, through use of his personalised GeAI EI coach, he transformed his communication style, listening more and offering empathetic feedback. His team, though initially sceptical, responded positively, and their faith in Mark and their working future together began to rekindle.

The process was long and challenging, but Mark evolved into a skilled leader. A leader that recognised his emotions, one that took a breath before speaking out and one that could tune into others' emotions and adapt, creating trustful and effective working relationships. Morale improved, productivity soared, and the company recovered. Mark's journey towards emotional intelligence continues, but he has successfully transitioned from a blunt engineer to a respected leader.

In Conclusion

The full extent of the capabilities and the associated opportunities of GenAI EI remains to be seen. Major advancements in GenAI model size, complexity, and efficiency appear every three to six months. Change is rapid, and while that is exciting and brings about bountiful opportunities, organisations must work phenomenally hard to keep up with data security and ethics.

Organisations need to be agile, always embracing innovation, seeking out, adopting, and adapting to this evolving technology. Employees too need to be open to this new dynamic, allowing for the technology to become a collaborative workplace partner that is necessary in order to progress and deliver business value and results. And like we have seen with many brands that fail to embrace innovation, it is a case of keep up or face decline and face the graveyard.

GenAI is not a fad. Its strength in terms of its intuition, knowledge, and accuracy is exponential. GenAI EI brings many positive opportunities for both organisations and employees, for personal growth and development, strengthening internal collaboration, and harmonising teams and leaders with their direct reports.

The above said, as always, the tech is merely a tool and should be applied as part of a multifaceted people strategy that incorporates the development of EI along with many other elements, such as wellness, diversity, and inclusion.

Although GenAI lacks true emotional experience, its immense knowledge and analytical prowess make it a powerful tool for improving emotional intelligence. By connecting the realms of logic and human feeling, we forge a special alliance: one where emotionless machines help us better understand our own emotions, at work and also in our personal lives.

References and Thank You

I'd like to thank Sandra Thompson of EI Evolution (https://www.eievolution.com) for opening my mind and heart to the concept of emotional intelligence during Covid, which led to my completion of my "applied CX and emotional intelligence" post grad.

Thank you also to Rich Hua, creator of EPIC for sharing with me the work done at both AWS and Splunk. https://aws.amazon.com/executive-insights/content/emotional-intelligence/

Press Release: More than 80 Per Cent of Enterprises to Adopt Some Form of Generative AI by 2026 (www.gartner.com)

https://www.superannotate.com/blog/llm-fine-tuning

About the Author

Kathryn Simons-Porter holds a master's in business management and a post grad diploma in applied CX and emotional intelligence. Starting her career initially in tech, selling contact centre solutions to some of the world's leading brands, Kathryn became a CX specialist in 2015 when she joined Merchants, a sister company of Dimension Data. Since, Kathryn has gone on to speak at a number of CX day events and is a member of various associations, including the global community, Women in CX.

A regular speaker at customer experience events, a CX awards judge, and frequent podcaster, Kathryn enjoys the psychology of human behaviour and more recently has examined workforce psychology and how technology, such as generative AI, can be part of a successful people strategy.

Kathryn currently is the Sales VP (EMEA) at Local Measure, a global Customer Experience technology company, working with some of the world's leading brands on their GenAI Contact Centre optimization programmes.

Outside of work, Kathryn enjoys a challenge in the name of charity, whether running a marathon, skydiving, or hiking. Kathryn lives in Yorkshire, England with her husband and dog, Willow.

Email: kathrynsp@flourishbusinessconsulting.co.uk

EMBRACING AI: HUMANITY'S PATH TO A TRANSFORMED FUTURE

By Thomas Somogyi
Head of Data, Automation and AI in IT Consulting
Zürich, Switzerland

AI is not just another tool in our arsenal; it is the very foundation upon which the future will be built, shaping every aspect of our lives and redefining the boundaries of what is possible.
—AI (ChatGPT)

Introduction: Setting the Stage

Artificial intelligence (AI) has been a subject of academic research since the mid-20th century, with early milestones such as Alan Turing's 1950 proposal of the Turing Test. Furthermore, the birth of AI as a field of study is often marked by the 1956 Dartmouth Conference, which was organized by John McCarthy and Marvin Minsky.

Over the past decade, however, AI has advanced rapidly, evolving from a developing field of theoretical exploration to a transformative force reshaping industries and societies worldwide. This recent surge in progress has been driven primarily by advances in computing power, the widespread availability of large datasets, and major breakthroughs in machine-learning techniques.

AI today has the potential to solve complex problems, drive innovation, and transform economies, which is why it is attracting so much attention and hype.

This chapter explores AI's dual nature as an opportunity and a risk and provides guidance for navigating this new landscape.

The Current vs. Potential State of AI

AI can be categorized by capability and function. Regarding capability, we have narrow AI, general AI, and superintelligent AI:

- *Narrow AI* is designed and used for specific tasks, such as voice assistants and virtual assistants, such as Siri and Alexa. It also includes autonomous vehicles, self-driving cars, and image recognition technologies in security and social media. In addition, advances in generative AI, also part of narrow AI, have facilitated the production of graphic art, music, and text, transforming content creation and design.

- *General AI*, which remains hypothetical, would have human-like intelligence and versatility, allowing it to perform any intellectual task that a human can. This level of AI would be able to understand, learn, and apply knowledge in a way that mirrors human cognitive processes, making it adaptable to a wide range of scenarios.

- *Superintelligent AI*, a theoretical construct, would exceed human intelligence across all domains. Such an AI would not only perform tasks faster and more efficiently than humans, but would also have the ability to innovate and solve complex problems that are currently beyond human comprehension.

Functionally, there are three main types of AI: reactive machines, limited memory systems, and theory of mind AI, each with different capabilities and applications:

- *Reactive machines* provide simple task-based responses without memory, such as IBM's Deep Blue, which defeated chess champion Garry Kasparov. These systems can analyze specific situations and respond appropriately, but they cannot use past experiences to influence future actions.

- *Limited memory systems*, such as self-driving cars, use past experiences to inform current decisions by recognizing and adapting to patterns. Such systems can store data temporarily to improve decision-making processes, allowing for more dynamic and context-aware responses.

- *Theory of mind AI*, still in the research phase, aims to understand human emotions and intentions, which would enable more sophisticated human-machine interactions. Theory of mind AI aims to build relationships based on empathy and social awareness, paving the way for more intuitive and effective collaboration with humans.

- *Self-aware AI*, as in AI possessing consciousness, remains speculative and is often the subject of philosophical and ethical debates. Achieving self-awareness in machines would require not only advanced cognitive capabilities, but also an understanding of subjective experience, a challenge that continues to drive significant discussion and research in the field.

Despite rapid advancements, AI's capabilities are still relatively basic compared to human intelligence. Think of AI as a friend who knows a lot of trivia but can't hold a deep conversation. Today's AI excels in pattern recognition and data processing but still needs to gain humans' general understanding and reasoning abilities. Misconceptions about AI often exaggerate its capabilities and potential threats, highlighting the need for clear, informed discourse.

AI—Humanity's Greatest Challenge

AI represents a significant turning point in human history, similar to previous industrial revolutions, but on a larger scale. The societal, ethical, and existential challenges posed by AI require careful consideration. Issues such as privacy, bias in AI algorithms, and the potential for autonomous error management require proactive oversight.

AI systems can have biased effects, particularly in areas such as hiring, lending, and law enforcement. In hiring processes, AI algorithms might unintentionally prioritize certain demographics, resulting in unfair selections. Privacy concerns are heightened as AI enables more sophisticated surveillance, raising fears about the erosion of personal privacy.

The ethical implications of AI are complex, particularly in addressing conflicts and societal challenges. Autonomous systems raise concerns about accountability and moral implications due to their ability to make decisions without human intervention. Therefore, to ensure the ethical development and use of AI, it is essential for the international community, including policymakers, international organizations, governments, and individuals involved in AI development and governance to collaborate on establishing rules and standards. Common standards and regulations can guide AI development according to ethical principles, minimizing risks and maximizing benefits for humanity. It's also crucial to recognize the critical role of each individual to ensure the responsible development and use of AI.

An Opportunity and a Major Risk

AI offers not just potential benefits but enormous benefits across multiple sectors. In healthcare, AI can improve diagnostics, customize treatment plans, and accelerate drug discovery. AI algorithms can enable accurate analysis of medical images, helping doctors detect diseases at an early stage.

Environmental conservation efforts can benefit from AI's ability to analyze large data sets, optimize resource management, and monitor ecological changes. Currently, AI-powered drones are used to track wildlife and monitor deforestation. Technological development driven by AI can lead to innovations that improve efficiency and productivity in manufacturing, logistics, and customer service.

However, these opportunities also come with significant risks. Job displacement due to automation is a primary concern, as AI and robots replace many traditional roles. In the future, autonomous vehicles could reduce the need for drivers, and AI-powered chatbots could replace customer service representatives. In addition, ethical concerns such as privacy and algorithmic bias must be carefully considered. Furthermore, the potential misuse of AI, from surveillance to autonomous systems, poses serious threats.

Economic disruption is a critical risk, with widespread automation potentially leading to significant job losses. Some countries have proposed universal basic income (UBI) to alleviate the economic impact and provide a safety net for those affected. UBI's feasibility, benefits, and potential drawbacks must be carefully considered. In addition, wage structures may need to evolve to accommodate a future where AI and humans coexist in the workforce, ensuring fair compensation and opportunity for all.

AI Should Enable Us to Be Better

AI has the power to enhance human capabilities, solve complex challenges, and improve the overall quality of life. Rather than replacing human skills, AI can collaborate and work alongside humans effectively. With an AI Mindset, humans will be able to work together with machines to create a positive outcome. Picture AI as having a super smart assistant who never needs a coffee break and constantly works alongside us to create a cooperative relationship between humans and machines.

AI can be a game changer in the fight against global problems such as climate change, poverty, and healthcare disparities. For example, AI can optimize energy use by analyzing patterns and predicting consumption needs, model climate scenarios to help policymakers plan for the future, and develop sustainable technologies such as smart grids and renewable energy solutions. In healthcare, AI can provide personalized medicine by analyzing genetic data, predict disease outbreaks by monitoring social media and other data sources, and improve patient care through AI-driven diagnostics and treatment recommendations.

In addition, the concept of spiral dynamics, a model of human development and consciousness, outlines multiple levels of increasing complexity and integration. This model complements the vision of achieving a type I civilization (explained below) by emphasizing the importance of self-awareness, continuous growth, and ethical evolution. By embracing these principles, we can ensure that AI enhances human potential and societal progress.

AI and the Next Level of Civilization

AI has the potential to propel humanity to the next stage of civilization, which aligns with the higher levels of the Kardashev scale. The Kardashev scale, proposed by Soviet astronomer Nikolai Kardashev in 1964, is a theoretical method of measuring a civilization's level of technological advancement based on its energy consumption. It consists of three primary stages:

- Type I Civilization: utilizes all available energy on its home planet

- Type II Civilization: harnesses the total energy of its home star

- Type III Civilization: controls energy on the scale of its entire galaxy

Humanity is not yet a type I civilization; we are estimated to be around 0.7 on the Kardashev scale, using a fraction of the energy available on Earth.

Critics of the Kardashev scale claim that it oversimplifies the complex nature of technological and societal progress and fails to consider sustainability and ethical implications. Since humanity has yet to reach these stages, the scale remains a theoretical benchmark for future advancements.

Achieving progress along the Kardashev scale requires the responsible and sustainable use of AI capabilities. By ethically leveraging AI's capabilities, humanity can unlock new frontiers in science, technology, and societal development, potentially advancing toward a type I civilization.

In the short term, we expect AI to continue improving efficiency and decision-making across industries. Long-term hypothetical scenarios for AI-enhanced societies could include improved resource management, extended human lifespans, and enhanced cognitive abilities. However, these advances must be pursued responsibly, ensuring equitable access and preventing misuse. Imagine a world where everyone, regardless of their background, can benefit from AI's capabilities.

The Benefits of AI Across Industries

As mentioned, AI's impact spans through multiple industries, driving efficiency and innovation. With an AI Mindset, humans will be able to improve efficiencies in many areas. Even though in healthcare, AI is improving diagnostics, personalizing treatments, and accelerating research, it's important to note that AI is not replacing doctors or researchers, but rather enhancing their capabilities.

The financial sector benefits from AI-driven risk assessment, fraud detection, and personalized banking services. AI algorithms can analyze financial data faster and more accurately than humans, identifying trends and spotting anomalies. Yet, in certain cases human oversight is crucial in these applications to ensure that AI is used ethically and responsibly.

In transportation, AI is revolutionizing the industry through autonomous vehicles, optimized logistics, and intelligent traffic management. AI can reduce traffic congestion by predicting patterns and suggesting the best routes, making commuting smoother and more efficient.

In education, AI personalizes learning experiences by providing tailored instruction and resources based on the student 's individual needs.

Collaborative AI-human projects have accelerated progress and solved complex problems. For example, AI-assisted research has led to breakthroughs in genomics, identifying genetic markers for diseases and potential treatments.

AI-driven data analytics supports critical decision-making in business and government by providing insights from vast amounts of data. Imagine this as having an analytics team that can analyze enormous amounts of data at incredible speeds to uncover insights that drive strategic decisions.

Future Potential with Emerging Technologies

Emerging technologies hold the promise of further advancing AI capabilities. In the short term, AI advances will continue to enhance existing technologies and applications. For example, AI will be further integrated with internet of things (IoT) devices, creating smarter homes and cities. Improvements in machine learning algorithms will enable more accurate predictive analytics and personalized experiences.

In the longer term, technologies such as quantum computing could revolutionize AI by exponentially increasing processing power and enabling the handling of complex calculations and large data sets. Quantum computers could simulate molecular interactions for drug discovery on an unprecedented scale. However, widespread practical applications may be years away. Nanotechnology could create new AI interfaces and capabilities, enhancing interaction and efficiency. Organic computing, which uses biological components, offers innovative approaches to AI development, potentially leading to more adaptive and efficient systems.

In addition, blockchain technology can play a critical role in ensuring transparency, security, and decentralized control in AI systems. Blockchain can improve the validation of AI decisions and maintain data integrity, reducing the risk of tampering and fraud. The integration of these technologies could lead to unprecedented advancements, driving progress in AI and beyond. Addressing ethical and security considerations in the development and deployment of these technologies alongside AI is essential to ensure that their benefits are realized responsibly. Think about giving the AI an extra power boost, without the capes and secret identities, to ensure its abilities are used for good.

Practical Advice and Ethical Considerations for Navigating AI's Disruption

To keep up in an AI-driven job market, individuals must focus on continuous learning and developing soft skills that AI cannot replicate, such as creativity, emotional intelligence, and critical thinking. Lifelong learning and adaptability are critical to thriving in this evolving landscape. Remember, while AI can beat you at chess, it still can't negotiate a car deal as well as you can.

Businesses must integrate AI ethically and effectively, ensuring transparency, fairness, and accountability. Corporate social responsibility is essential to contribute to a positive AI-driven future. Companies should also invest in reskilling their workforce, promoting diversity in AI development, and prioritizing ethical AI practices. This includes developing AI that is accountable (protocol-driven) and fair and engaging diverse teams in the development process to avoid bias.

Tackling the challenges of AI requires collaboration across disciplines. Experts in computer science, ethics, sociology, and law must join forces to develop well-rounded solutions. Ethical frameworks and regulations should guide the development and use of AI, balancing innovation with the protection of human values.

Conclusion

In conclusion, AI presents humanity with unprecedented opportunities and significant risks. By taking a balanced approach, focusing on ethical considerations, and promoting continuous learning, we can maximize the benefits of AI while mitigating its potential drawbacks. A proactive and thoughtful approach to AI developments will help shape a positive future and ensure that AI serves as a force for good in advancing civilization. Think of it as navigating uncharted waters with a high-tech compass—exciting but requiring careful steering.

Individual Tips for Benefiting from AI and Creating an AI Mindset

- *Openness and Willingness to Learn*: stay informed about AI advances and their impact on your field. Embrace new technologies and be open to change.

- *Continuous Learning*: invest in lifelong learning to acquire new skills that complement AI, such as data analysis, programming, and critical thinking.

- *Curiosity and Experimentation*: experiment with AI tools and applications to understand their potential and limitations. Use AI to increase your productivity and creativity.

- *Networking and Collaboration*: connect with professionals in the AI community to share knowledge, ideas, and opportunities. Collaborate on projects that use AI to solve real-world problems.

- *Ethical Awareness*: be mindful of the ethical implications of AI in your work. Advocate for transparency, fairness, and accountability in AI systems and practices.

About the Author

Thomas Somogyi is an experienced business and IT professional with 18 years of experience in project management, IT, and business consulting. He holds degrees in software development and business informatics, as well as an MBA and an EMBA in business psychology. His diverse career began with service in the army, where he learned valuable lessons in discipline, resilience, and the importance of teamwork.

After his military service, he explored different roles, working in security, sales, customer service for entertainment and IT products, remodeling houses, moving services, professional cleaning, and as a promoter for non-governmental organizations (NGOs). Trying out different jobs has given him a broad insight into different industries and perspectives.

In Thomas's day-to-day work, he uses AI to enhance his capabilities and speed in software development, data and text analysis, and ensuring concise communication. He also leverages AI for brainstorming, writing out ideas, and organizing thoughts with a virtual assistant while double-checking and verifying the AI content to ensure the validity of information and copyright compliance. He utilizes different AI solutions, from public ones to closed environments and protected tenants, depending on the sensitivity of the information.

During his free time, Thomas enjoys traveling and meeting people around the world, which enriches his cultural understanding. He experiences the beauty of multiculturalism every day with his wife, who is Argentinian, Chilean, and American. He also practices martial arts, which contributes to his discipline and physical well-being. He describes himself as cosmopolitan and philanthropic, striving to make the world a better place for everyone, including animals and nature.

Thomas believes in approaches such as "in fidem," "carpe diem," and "we live and learn," emphasizing self-awareness, continuous growth, balance, and connection to the cosmos and the energy that connects us all.

E-Mail: info@thomassomogyi.com
Website: www.thomassomogyi.com

SECURING THE FUTURE: THE ROLE OF AI IN CYBERSECURITY

By Dan Sorensen
USAF/ANG Veteran, CISO, Cybersecurity Engineer
Denver, Colorado

I am super optimistic about the near-term prospects of AI because every time there is a technological disruption, it gives us the opportunity of making the world a little different.
—Andrew Ng

AI is transforming industries at an unprecedented pace. Just as electricity revolutionized manufacturing and technology, AI is now reshaping fields like healthcare, finance, and, perhaps most critically, cybersecurity. As cyberattacks become more frequent and sophisticated, traditional defense mechanisms can no longer keep up. AI offers advanced detection, response, and remediation capabilities, allowing cybersecurity teams to respond at a scale and speed that surpasses human limitations. This chapter explores the evolving intersection between AI and cybersecurity, detailing the opportunities AI provides and the challenges we must address to secure the digital future.

The Current Cybersecurity Landscape

Over the past decade, the cybersecurity landscape has changed dramatically. According to the Open Web Application Security Project (OWASP) AI Exchange and the European Union Agency for Cybersecurity's (ENISA) directives, cyberattacks are not only increasing in frequency, they are also growing more sophisticated, costing organizations billions of dollars in losses annually. Governments and businesses are vulnerable to ransomware and state-sponsored advanced persistent threats (APTs).

The rise of remote work, cloud infrastructure, and internet of things (IoT) devices have made traditional perimeter-based defense strategies insufficient. As noted by the ENISA, it is now crucial to integrate AI into modern cybersecurity frameworks to effectively combat today's complex threats.

AI in Cybersecurity: An Overview

AI's capabilities in cybersecurity are powered by machine learning (ML) and deep learning (DL) algorithms. These technologies analyze vast amounts of data, identify anomalies, and predict threats, allowing real-time threat detection and response. ML algorithms can analyze historical data to identify malicious patterns in network traffic and behaviors, enabling faster and more accurate detection of threats. This advanced form of ML allows AI systems to process more complex data, such as images or documents, to detect threats embedded in unexpected places, like hidden malware in PDFs or phishing attempts within images.

Historically, cybersecurity relied on static signature-based systems that struggled to adapt to evolving threats. AI, however, adapts and learns from the heuristic behavior of an organization's network and each new attack vector, making it an indispensable tool for modern defense strategies.

Applications of AI in Cybersecurity

AI-powered tools can analyze real-time large-scale network traffic, flagging suspicious activities based on behavior and anomaly detection algorithms. A significant healthcare institution utilized AI-driven intrusion detection systems (IDS) to identify abnormal data transfers within its internal servers. The IDS flagged the behavior, which was revealed as an insider threat attempting to steal patient data. Traditional systems may have missed this, but AI's ability to link subtle indicators allowed for early detection.

AI can also streamline incident response processes. AI-driven systems can automatically assess the severity of threats, recommend countermeasures, and initiate responses like isolating malware or alerting security teams. Security information and event management (SIEM) tools integrated with AI-powered security orchestration, automation, and response (SOAR) platforms can detect and isolate compromised systems in real time, drastically reducing incident response times from hours and days to only minutes.

AI can predict potential vulnerabilities based on historical usage and attack patterns, allowing organizations to patch systems before exploits become active. AI-powered vulnerability management ensures systems are proactively protected. AI enhances biometric and behavioral authentication by analyzing typing patterns or login times. Behavioral analytics, powered by AI, help detect anomalies and prevent unauthorized access to sensitive systems.

AI-Driven Cybersecurity Tools

Many organizations have adopted AI-driven cybersecurity tools to boost their defense capabilities. Some popular tools include integration into firewalls, IDS, intrusion protection systems (IPS), and SIEM/SOAR platforms. These firewalls use AI to predict and block malicious traffic based on past behaviors, industry-specific attacks, and tactics, techniques, and procedures (TTPs) of APTs of known script kiddies and nation-state cyber actors rather than relying solely on predefined rules. AI-powered IDS/IPS can detect and alert teams to suspicious network activity in real time, mitigating threats before they escalate at higher detection rates than typical IDSs with high false positives and false negatives. Adding AI capability can reduce false positives and provide real-time alerts about suspicious activity across multiple networks worldwide. Combining AI with automation, SOAR platforms provide unified responses to threats across an organization's entire security ecosystem rather than responding in isolated silos across on-premises, cloud-based enterprises, and large-scale subnets.

While these tools provide enormous benefits, they also present challenges, including introducing biases within AI models and the potential for missed or misinterpreted threats. AI cannot fully replace the human analyst's role but can assist at larger scales.

Challenges and Risks of AI in Cybersecurity

While AI benefits cybersecurity, it presents unique risks, such as AI-powered ransomware or botnet attacks. Cybercriminals increasingly use AI to develop adaptive malware capable of learning from security systems. These AI-driven attacks can evade detection by altering their behavior in response to initial reconnaissance. AI systems are only as good as the data they are trained on. The AI may fail to recognize specific threats or over-flag benign behaviors if the data is incomplete or biased. Ensuring ethical AI use is crucial to avoid disproportionate security responses. AI systems rely heavily on data collection, raising significant privacy concerns. Regulations like the General Data Protection Regulation (GDPR) mandate transparency and accountability regarding how AI collects and processes personal data. AI's decision-making processes can be opaque, which makes it difficult for organizations to trust these systems in critical situations. AI models must be transparent and explainable, particularly in environments where human lives or sensitive data are at stake. The 2023 US White House Executive Order on AI calls for mandatory safety testing of powerful AI models, ensuring transparency and safety.

Future Trends and Innovations

As AI continues to evolve, so too will its applications in cybersecurity. Future innovations will likely include AI-powered quantum cryptography, predictive cyber defenses, and integration into zero-trust architecture. AI will likely play a role in developing quantum cryptography, creating encryption methods that are far more advanced than today's standards. AI will increasingly be used to help predict cyberattacks by analyzing information sharing and analysis centers (ISACs), the MITRE ATT&CK Model fused with global cyber threat intelligence (CTI) portals, and anticipating trends from new and emerging threats on VirusTotal, AT&T's Alien Lab Open Threat Exchange (OTX), and Recorded Future. This current emerging trend is forecasted to see AI at the core of "zero trust" security models. Although not currently "zero trust," as it can verify every user and device before granting access to systems, the AI systems will allow it to associate disparate data and cyber-attacks faster than before.

Best Practices for Integrating AI in Cybersecurity

Organizations should adopt AI strategically to maximize its effectiveness, just as it was necessary to adopt networks, WIFI, and the dot-com boom. Some of these best practices include training security teams on the proper use of AI, using AI in continuous monitoring (CONMON), and ensuring the ethical use of AI. Cyber defense teams must understand how to operate AI-driven tools effectively and ethically. AI should assist, not replace, human analysts. These tools, whether driven by automation, ML, or generative AI, are just that; they assist in the human analytical process as a cybersecurity professional, but they do not replace people's critical judgments and instinct. AI can help pick out and showcase specific analytical data in near real time that otherwise would not be available for a cybersecurity analyst to extract from seemingly disparate pieces of data quickly.

As with any tool, AI should be paired with human oversight, ensuring that decisions made by AI systems are validated and understood, not taken out of context or given the whole choice as to whether a network is considered "secure" or insecure based solely on one analytical assistive tool. Organizations should adopt AI within ethical frameworks, ensuring compliance with guidelines, principles, frameworks, and laws like those of the Institute for Ethical AI & Machine Learning, the Responsible AI Institute, the European Union's (EU) GDPR, and the California Consumer Privacy Act (CCPA), and they should consider the long-term societal impact of AI deployment.

Limitations of Using AI in Cybersecurity

AI technologies like ML, generative pre-trained transformers (GPTs), large language models (LLMs), and generative AI (GenAI) offer significant potential but also face limitations, mainly when applied to air-gapped systems (systems that are isolated from external networks), industrial control systems (ICS) in factories and power plants, and supervisory control and data acquisition (SCADA) systems that monitor and control systems like pipelines or electrical grids from a central location.

ML models require vast amounts of data, which may not be available in air-gapped or standalone systems. This can be problematic in cybersecurity, as attack data is often rare, unbalanced (with far more "normal" network data than attack data), and rapidly evolving. There are knowledge cutoffs on many GPT models, such as GPT-4, which cannot be updated in real time, making them less effective in dynamic threat landscapes. If the training data does not represent all possible scenarios (e.g., zero-day attacks where no known defenses have been established yet), there can be training bias, and ML systems can misclassify new or evolving threats as "false negatives." ML models are prone to generating false positives, as typical activity is wrong, or false negatives of nefarious activity as typical day-to-day operations. In cybersecurity, this can lead to missed attacks or unnecessary alerts, which can overwhelm security teams. Attackers need to get it right once, whereas cyber defenders need to get it right every time to effectively defend.

For air-gapped systems or systems that operate independently—typically the systems operating electric grids, water treatment facilities, and other extensive infrastructure systems—the lack of access to real-time threat intelligence or diverse training data limits the ability of ML models to adapt to emerging threats. These systems often operate on older, proprietary protocols with minimal changes, making it difficult for ML models to adapt. Additionally, the risk of disrupting mission-critical operations makes using predictive models and automated responses risky.

API integration into GPT models like GPT-3 or GPT-4 are trained on vast datasets but have a knowledge cutoff (i.e., they typically do not update in real time). In cybersecurity, where threat landscapes evolve daily, this lack of up-to-date information can render GPTs less valid for real-time threat detection or mitigation. Some previous GPT frameworks and models have been released on Hugging Face and similar open-source forums but are not as complex as the newly released GPTs. GPTs generate text based on patterns but often lack contextual understanding of ongoing cybersecurity incidents. This can result in outdated, generic, or incomplete responses in high-stakes situations.

GPT models typically require connection to external systems and/or the internet for training or real-time inference. The lack of internet or external data sources severely limits the utility of trained GPT-based solutions and updated holistic integrations for air-gapped or standalone systems. GPTs are text-based models and cannot directly interact with the low-level, real-time data streams standard in ICS/SCADA environments.

These systems' unique protocols and operational requirements are unsuited to small or large language-based models.

Retrieval-augmented generation (RAG), an AI method that retrieves relevant data from a source and then uses a language model to generate a response based on that information, is part of the process that can help AI provide more accurate and up-to-date answers. These RAG systems also integrate external knowledge bases to improve the accuracy and speed of computation. In cybersecurity, threats' dynamic and complex nature requires constantly updated external knowledge bases, which might not always be available or reliable, or they are over-reliant on biased features such as certain types of cyber-attacks, but not others.

RAG systems, especially more complex trained ones, can introduce latency or inconsistencies due to the retrieval and generation steps, which can be problematic in cybersecurity, where real-time response is crucial. Since RAG models need to retrieve external information to enhance their responses, they cannot function as effectively in isolated environments like air-gapped systems, which have no access to the internet or external data sources. The latency and complexity of RAG make it unsuitable for real-time operational environments like ICS and SCADA, where immediate responses are necessary to maintain safety and operations.

LLMs typically operate as "black boxes," which makes it difficult for programmers and cybersecurity professionals to understand the logic behind their recommendations. This is problematic in environments where accountability and transparency are critical. LLMs trained on sensitive data can inadvertently expose confidential information. In cybersecurity, where protecting data is paramount, this represents a significant risk. Like GPTs and RAG, LLMs rely on internet connectivity and ongoing updates of data points for their operation. These models are not feasible for air-gapped systems operating independently of external data sources. LLMs are not designed to interact with the highly specialized and real-time data requirements of ICS/SCADA systems. LLMs' reliance on general-purpose training data limits their applicability to niche, mission-critical environments.

GenAI models can generate inaccurate or misleading information if prompted incorrectly or trained on biased data. In cybersecurity, incorrect responses can lead to poor decision-making, exacerbating security risks and producing unintended results or confusing disparate analytic results. GenAI models can be targeted by adversarial attacks, where subtle manipulations in input data cause the model to generate harmful or incorrect outputs, which do not necessarily change the program's overall functionality as if it were software but cause it to accept false data as valid data points. GenAI models require significant computational resources and constant updates from external sources, making them unsuitable for isolated systems. The deterministic nature of ICS/ SCADA systems, where uptime and reliability are paramount, contrasts with the probabilistic nature of generative models. As a result, GenAI is less effective in environments where strict control and predictability are necessary.

Air-gapped and standalone systems, often used in military, critical infrastructure, or sensitive corporate environments, are isolated from external networks to minimize the risk of external attacks. This isolation poses challenges for AI-driven cybersecurity solutions that rely on real-time data, threat intelligence, and regular model updates. AI models must operate with limited datasets, limiting their ability to identify new or evolving threats to the ecosystem, industry, or world. ICS and SCADA systems often run on legacy software and hardware that cannot be easily integrated with modern AI models. These systems also have low downtime tolerance, which makes it challenging to implement AI-based systems requiring frequent training, updates, or adjustments.

Lack of Specialized Models

AI models, especially those designed for general-purpose tasks like LLMs or GPTs must be better suited to the specialized protocols and environments of ICS/SCADA systems. These systems require precise, real-time controls, while AI models, particularly those in cybersecurity, often operate in probabilistic terms and with less predictable outputs.

In environments like ICS and SCADA, real-time decision-making is critical. Many AI systems, especially those based on deep learning, require time to process and analyze data, which can introduce unacceptable delays in systems where split-second decisions can impact safety or operations.

Human-in-the-Loop Limitations

AI models as a tool to help human analysis often require human oversight, particularly in sensitive environments like ICS or air-gapped systems. However, integrating AI into these systems may overwhelm human operators with data or false alerts, reducing the effectiveness of human-in-the-loop systems and potentially causing operator fatigue.

Ethical and Legal Constraints

AI in cybersecurity must comply with stringent regulations, especially in critical sectors like energy, healthcare, and defense. The European Union's General Data Protection Regulation (GDPR) and the Cybersecurity Information Sharing Act (CISA) impose limitations on how AI can collect, process, and store data, particularly sensitive or personal information. AI models are only as good as the data they are trained on. AI models trained on biased or insufficient data can introduce vulnerabilities in environments like ICS or air-gapped systems. Moreover, there is an ethical concern about over-relying on AI for decision-making, mainly when human lives or critical operations are at stake.

Conclusion

AI can potentially transform cybersecurity in ways we are only beginning to understand. It can enhance our defenses, predict threats, and automate responses with unprecedented accuracy. However, to secure our future, we must address the ethical and operational challenges AI introduces. AI offers powerful "assistive" tools for enhancing cybersecurity, but its limitations, mainly when applied to air-gapped, standalone systems, ICS, and SCADA, must be acknowledged. The complexities of these systems—ranging from real-time requirements to limited connectivity— pose unique challenges that AI technologies such as ML, GPTs, LLMs, and GenAI are not always equipped to address effectively.

The world of cybersecurity will continue to evolve alongside AI, and embracing this technology ethically and effectively is critical to building a safer digital landscape. To maximize the benefits of AI in these environments, organizations should combine AI solutions with human oversight, continuous monitoring, and well-established cybersecurity protocols to mitigate risks and ensure the reliability of critical systems.

About the Author

Dan Sorensen is a seasoned cybersecurity leader with over 22 years of experience as a CISO and cybersecurity engineer in aerospace. A US Air Force and Air National Guard veteran, he specializes in risk management, AI-driven defense strategies, and ethical AI integration. Dan has worked globally with military, governmental, and corporate sectors, securing critical infrastructure against advanced threats.

Dan is a recognized thought leader and a distinguished listee in Marquis Who's Who. He frequently speaks at industry conferences on AI, cybersecurity leadership, and the future of digital defense. Dan serves as the CSFI Defensive Cyberspace Operations Board of Education Chair and is a member of the Forbes Technology Council, Fast Company Executive Council, the Institute for Ethical AI & Machine Learning, Responsible AI, and the Air Force Association (AFA) Mile High Chapter of Colorado. He is also an active contributor to the OWASP AI Exchange Core Team.

LinkedIn: https://www.linkedin.com/in/cisoDan/

ENHANCING PRODUCTIVITY AND MOTIVATION WITH AI

By Sakina Syed, B.Sc.
Microsoft IT Professional, AI Engineer and Consultant
Toronto, Ontario, Canada

*AI is the most important thing humanity has ever worked on.
I think of it as something more profound than electricity or fire.*
—Sundar Pichai

In today's fast-paced work environment, boosting productivity is vital for businesses to remain competitive. One of the most effective ways to achieve this is by having an improved cohesive outlook on business processes, automating repetitive tasks, utilizing data to gain better insights, applying intuition to gain a holistic perspective, and predicting ways to optimize various business areas. In comes AI with the superpower-like ability to solve this challenge. AI can manage mundane tasks and provide better insights into how we can be more efficient, allowing us to do much more and become more productive individuals.

Whether it's helping us get more work done in less time, predicting the weather more accurately, finding patterns in datasets and spreadsheets, and creating a new way of approaching problems with these tools, AI is going to change the way we work forever. AI can learn to recognize things to do and extract insights faster than the average human can. These tools can also help us solve complex problems that we might not be able to on our own. AI is reimagining the way we work. Think of it as a smart brain for computers and a tool that humans can leverage to become super-productive versions of themselves. In this chapter, we will explore how AI is empowering organizations to achieve more and allow their employees to be more motivated.

AI's Role in Enhancing Productivity in an Organization

Fifteen years ago, AI was in its infancy, limited by computational power, data availability, and its inability to handle new and unseen tasks. No one could have predicted its current advancement. Today, thanks to improvements in computational power, vast data availability, and algorithmic maturity, AI can manage complex tasks and even those that it has never encountered before with added intuition. AI now enhances productivity not just by performing routine tasks but also by learning and adapting to new ones on the fly, making it an indispensable tool in today's business landscape.

AI's Role at Enhancing Operational Efficiency

Thanks to AI, streamlining processes and reducing the time required to complete tasks has become easier than ever before. This leads to an increase in operational efficiency across organizations in many fields. For instance, let's think about organizations that need to respond to customers at late hours, but they don't have the manpower to keep their operations running past business hours. AI-powered chatbots can be the solution to handle customer inquiries around the clock while reducing wait times and improving customer satisfaction, and allowing company employees to work in the daytime without burnout.

Automating Repetitive Tasks

It's rare to like doing the same task over and over and over again, such as data entry, taking meeting notes, scheduling meetings, or managing emails. These tasks can feel redundant and cause mental drain for employees, so why not offload them to AI and allow members of an organization to move on to more meaningful work? AI uses algorithms that follow your prompts, and you can command it to handle tedious and mundane activities. For example, you can direct AI tools such as Copilot with the following prompt: "Summarize the transcribed notes from the Teams meeting today on our research proposal."

Why should you consider allowing AI to do this? AI excels at handling routine and repetitive tasks. By automating these tasks, employees are freed to focus on complex and creative work. This not only increases productivity but also enhances job satisfaction as employees can engage in more significant activities. AI can be used in content moderation and filtering, customer support, email management, scheduling, calendar management, note-taking and transcription, R&D, and the list goes on.

Another example in the case of automating tasks is an AI-powered tool known as robotic process automation (RPA), which can process invoices, update customer records, and manage inventory, thus significantly reducing the time spent on these tasks. SAP, the software company, implements RPA for financial organizations to automate complex financial processes. Automation also reduces the probability and risk of human errors, which leads to a more reliable and consistent operational environment and, of course, as a result, an increase in productivity.

What Is AI's Role in Enhancing Decision-Making?

Well thought-out decision-making at the right time can propel an organization years ahead while a decision at a critical moment in business can make or break an organization's success. An AI assistant can help business decision makers and employees make better and more informed decisions more quickly than ever before, without limitations. AI can analyze vast amounts of data to identify trends, patterns, and insights that might not be immediately apparent to humans. This is thanks to LLMs or large language models that can analyze and generate insights from text prompts, documents, Excel spreadsheets, and other types of data.

We can think about AI tools such as Microsoft Copilot for M365, which is based off of Open AI's Chat GPT-4 LLMs, which integrates its intelligent text analytical capabilities to be able to take meeting notes and analyze them to extract further insights, sort through email threads and derive action items, generate PowerPoint slideshows and content in Word documents, and, thus, allow employees to do their best work in less time. These AI models are designed to understand and generate human-like text based on the input they receive. This capability is invaluable for leaders who need to make strategic decisions while also having their decisions backed by data.

For instance, let's think about a business decision maker, who in a board meeting needs to present a plan of action for how the company aims to improve its performance against competing products. AI algorithms can analyze market trends, gather perspectives discussed in meetings, make an informed recommendation or suggestion, and also analyze customer behavior and financial data to provide actionable insights. This helps organizations make data-driven decisions, reducing the risk of errors and improving overall efficiency.

Think of AI as a personal assistant to help you move in the right direction, with one more added benefit: AI doesn't get tired and it is there as a tool to help you achieve more, whenever you need it.

Optimizing Operations

If there was a machine invented to gain back more time, everyone would want one. Businesses everywhere have a need to save time and improve their efficiency, accuracy, and overall operational performance. While AI can't allow you to time travel, it does give employees back more time in their day by streamlining processes, which allows them to be more productive. Let's take a closer look at this.

AI can streamline various operational processes. Let's think about an AI system that can conduct quality control. AI-powered systems such as Pipefy can monitor production and detect defects, which ensures higher-quality products and reduces waste. This also creates more long-term sustainability for both businesses and the environment.

Across supply chain operations, AI can optimize supply chains and predict demand and maintenance needs. Predictive maintenance powered by AI can forecast equipment failures before they occur, minimizing downtime and maintenance costs. This is achieved in conjunction with IoT (internet of things) devices that provide real-time data to these integrated AI-driven solutions. IBM Watson is a leader in this segment. AI is also great at analyzing production data workflows in real time to identify bottlenecks and suggest improvements, ensuring smoother operations.

Boosting Employee Productivity

Everyone needs a boost now and then. Whether it is addressing writer's block by generating outlines for documents and presentations, image generation, coding guidance, or helping to craft an email, AI is at your service with the availability of a range of AI assistants to choose from.

AI tools can also assist employees by providing recommendations and even offer training and development opportunities. For instance, AI-driven platforms can suggest the best ways to complete tasks, recommend learning resources, and even provide personalized training programs, keeping the employees' needs in mind along with accessibility. This helps employees improve their skills and productivity, contributing to the overall growth of the organization.

Bored or Burnt Out? Leverage AI to Enhance Creativity

In today's era of hybrid work and back-to-back meetings, employees are more prone to burnout and often struggle to complete their work in normal hours, leading to a lack of innovation and time invested in thinking of new ideas. By automating repetitive tasks, AI allows employees to focus on innovation and creative problem-solving. For instance, AI can take over the task of taking meeting notes, freeing employees to engage more actively in discussions. With more time and mental space, employees can brainstorm new ideas, develop innovative solutions, and drive organizational growth. AI also aids the creative process by generating ideas, creating content, and even designing products, providing a valuable tool for creative professionals.

Consider the water treatment industry, where AI is revolutionizing processes by enhancing efficiency, accuracy, and sustainability. Christian Koppl, director of IT innovation at Ovivo, a leading company in water treatment solutions, shares insights on how AI influences employees' creative output and its applications across various industries:

> Effective communication skills are crucial for enhancing creativity and achieving work-life balance, with tools like Copilot aiding in this process. AI, while not yet capable of human-like contemplation, restructures jobs rather than eliminating them, emphasizing the need for continuous employee development. The ethics of AI emphasize the responsibility of corporations and businesses to

ensure their employees have opportunities for development and growth, such as returning to school on the company's dime.

As technology advances, it is crucial for employees to have the holistic skill set to identify and address issues when systems break down. Ensuring a seamless integration of AI into various roles can assist employees tremendously in gaining additional perspectives. In our own industry Microsoft Copilot was integrated and we have already seen many successful use cases for this tool as we embark on our AI transformation journey. The integration of AI into organizations will lead to a shift in focus, encouraging creativity and innovation.

AI can also enhance creativity by generating images and designs based on user prompts, as seen with generative AI tools like DALL-E. By providing detailed prompts, users can achieve precise results, making complex tasks like creating sculptures with 3D printers more efficient. We can be more articulate in the way we convey prompts and thoughts to AI tools such as the creative tool DALLE-E and provide it the context that it needs to generate creative possibilities such as images, designs, or with presentations of text. It's important to note that AI won't venture outside of our parameters of what we want to create, and the more information we give, the better the result or render. AI doesn't limit creativity but rather facilitates bringing ideas to life quickly and effectively.

Improving Motivation with Artificial Intelligence

Imagine if AI was able to project your career growth and help you stay motivated to reach it. AI has the potential to significantly enhance an individual's motivation by providing tailored support and feedback.

AI can achieve this through personalized learning and development. By analyzing an individual's strengths, weaknesses, and learning preferences, AI can create customized learning paths that align with their career goals and personal interests. A personalized approach ensures that an employee is engaged and motivated to learn because they can see a clear connection between their efforts and their professional growth.

From the Work Trend Index report study, "This new generation of AI will remove the drudgery of work and unleash creativity," said Satya Nadella, Chairman and CEO, Microsoft. "There's an enormous opportunity for AI-powered tools to help alleviate digital debt, build AI aptitude, and empower employees."

Consistent feedback helps employees thrive. AI can significantly enhance employee motivation by providing real-time performance feedback, which traditional reviews often lack. This immediate, proactive feedback loop allows an employee to adjust and make improvements on the fly, fostering a sense of progress and achievement. Additionally, AI plays a crucial role in recognizing and rewarding high performers, boosting morale and motivation. For those underperforming, AI is here to help them to meet their goals.

By leveraging these AI capabilities, organizations can create a more engaged, productive, and motivated workforce.

AI Support in Decision-Making

Artificial intelligence is revolutionizing decision-making processes across various industries by providing data-driven insights and predictive analytics.

One example is John Deere's application of AI and machine learning in precision farming. John Deere uses AI to analyze data from sensors placed in agricultural equipment and fields. This data includes soil conditions, weather patterns, and crop health. By processing this information, AI can provide farmers with actionable insights on the best times to plant, irrigate, and harvest crops. This data-driven approach helps farmers make decisions that maximize yield and minimize resource use, which leads to more sustainable farming practices.

In the automotive industry, AI is being used to support safe driving decisions. Companies like Tesla utilize AI to analyze data from vehicle sensors and cameras to make real-time decisions about navigation and obstacle avoidance. This technology not only improves the safety of autonomous vehicles but also provides valuable insights for future advancements in automotive technology.

Lastly, AI is transforming business strategies by enabling more informed decision-making processes. A study by the MIT Initiative on the Digital Economy highlights how AI can improve human decision-making by enhancing heuristics and everyday practices. By providing data-driven insights and predictive analytics, AI helps business leaders make strategic decisions that are more accurate and effective. This case study underscores the potential of AI to revolutionize decision-making across various sectors. The question to ask is whether this will take the power away from us to make decisions for ourselves, but like any tool, it's meant to enhance and add on to what we can do.

How to Use AI to Foster Innovation in Organizations

If an organization wants to take advantage of tapping into a continuous stream of new, creative ideas and innovation, they should leverage AI to stay ahead. AI can assist in ideating and generating new solutions derived from vast data sets and previous successful innovations. This ability allows teams to explore a wider range of options and solutions and find unique ideas that might not have been considered otherwise. AI is playing a vital role in fostering innovation by enhancing idea generation and accelerating research and development processes.

Think of the use case of AI analysis tools, such as Autodesk AI being integrated into software such as Adobe's AutoCAD. With the help of AI, architectural and engineering drawings and images can now be analyzed to create improved versions to meet compliance and safety standards and allow for better innovation. This is still an emerging technology, and it will continue to expand in capabilities.

Within the realm of research and development (R&D), AI notably accelerates processes by analyzing large volumes of data and identifying patterns and relationships that can lead to breakthroughs in innovation. To illustrate, AI can process scientific literature, experimental data, and patent information to uncover trends and insights that inform the development of new products and technologies. This ability to quickly synthesize and analyze information enables researchers to make informed decisions faster, reducing the time and cost associated with R&D and allowing for an overall increased level of productivity. As a result, AI-driven R&D can lead to more rapid advancements and provide organizations who leverage AI with a competitive edge in the market.

Ensuring Well-Being

AI is playing a crucial role in ensuring well-being by providing innovative solutions for mental health support and work-life balance.

For example, AI-powered chatbots and virtual counseling services are becoming increasingly popular for mental health support. These tools offer immediate, accessible, and often anonymous assistance to individuals experiencing stress whether it be short- or long-term, anxiety, or other mental health needs. For instance, AI chatbots like Woebot and Wysa use cognitive behavioral therapy (CBT) techniques to help users manage their mental well-being by providing real-time conversations and personalized support that is tailored to their needs. This approach not only makes mental health resources more accessible but also reduces the stigma associated with seeking help.

In addition to mental health support, AI is enhancing work-life balance by optimizing schedules and workloads. AI-driven tools can analyze employees' work patterns and suggest adjustments to prevent burnout and improve productivity. For example, AI can recommend optimal times for breaks, prioritize tasks, and even automate routine activities, allowing employees to focus on more meaningful work.

A notable case study highlighting AI's impact on well-being is the implementation of AI tools by the global consulting firm Accenture. Accenture uses AI to support its employees' mental health and work-life balance through a comprehensive wellness program. The program includes an AI-powered chatbot that provides mental health support and resources, as well as AI-driven scheduling tools that help employees manage their workloads more efficiently. This initiative has led to significant improvements in employee satisfaction and productivity, demonstrating the potential of AI to enhance well-being in the workplace. This technology helps bridge the gap in mental healthcare and makes getting support much easier.

The future of AI in assisting humans on a daily basis promises to be bright. Tesla is also in the process of actively training its humanoid robot, Optimus, also known as the Tesla bot, which can have a plethora of applications in the assisted living sector. While this may take some time to see a robot that can actually complete tasks in a human-like way, if successful, it will be revolutionary. Boston dynamics, a robotics design company, is also creating their own humanoid robot, Atlas, to complete mundane activities. With its athletic intelligence, dynamic manipulation, real-time perception, and model-predictive control it has the capabilities to be mobile and have dexterity and perception to bring out performance.

Conclusion

Overall, we looked at how AI is able to create space for a more productive and motivated workforce across organizations, and the outlook for adopting AI solutions is optimistic. By helping employees manage their time more effectively, AI contributes to a healthier work-life balance, which is essential for overall well-being. By adopting AI, organizations can spark creativity and enhance productivity for all, leading to a new era of growth and value creation.

References

Edelman. (2019). *2019 Edelman trust barometer special report: In brands we trust?* Edelman. https://www.edelman.com/research/trust-barometer-special-report-in-brands-we-trust

OpenAI. (n.d.). *DALL·E: Creating images from text.* OpenAI. https://openai.com/index/dall-e/

Microsoft. (n.d.). Learn about Microsoft Copilot. Retrieved October 30, 2024, from https://www.microsoft.com/en-us/microsoft-copilot/learn/?form=MA13FV&msockid=0b924a73ce5a64082c885967cff0650a

V7 Labs. (2023, October 30). *Prompt engineering guide.* V7 Labs. https://www.v7labs.com/blog/prompt-engineering-guide

Microsoft. (2023). *Work Trend Index 2023 Annual Report: Will AI fix work?* https://www.microsoft.com

Marr, B. (2019, March 15). *The amazing ways John Deere uses AI and machine vision to help feed 10 billion people.* Forbes. https://www.forbes.com/sites/bernardmarr/2019/03/15/the-amazing-ways-john-deere-uses-ai-and-machine-vision-to-help-feed-10-billion-people/

MIT Initiative on the Digital Economy. (2019, March 15). *How AI can improve human decision making.* Medium. https://medium.com/mit-initiative-on-the-digital-economy/how-ai-can-improve-human-decision-making-f70964659aae

Autodesk. (2024). *Autodesk AI in AutoCAD: Smart Blocks and Markup Import Assist.* Autodesk University. https://www.autodesk.com/autodesk-university/class/Autodesk-AI-AutoCAD-Smart-Blocks-and-Markup-Import-Assist-202

TechCult. (2023, October 30). *Best AI mental health chatbots.* TechCult. https://techcult.com/best-ai-mental-health-chatbots/

Healthline. (2023, October 30). *Chatbots for mental health: Reviews and insights*. Healthline. https://www.healthline.com/health/mental-health/chatbots-reviews#Wysa

Government Technology. (2024, October 31). How independent is Boston Dynamics' humanoid robot. *Government Technology*. https://www.govtech.com/question-of-the-day/how-independent-is-boston-dynamics-humanoid-robot

Innovation, Science and Economic Development Canada. (2023). *Artificial Intelligence and Data Act (AIDA) companion document*. Government of Canada. Retrieved from https://ised-isde.canada.ca/site/innovation-better-canada/en/artificial-intelligence-and-data-act-aida-companion-document

Nothing But AI. (n.d.). *H&M artificial intelligence*. Nothing But AI. Retrieved from https://nothingbutai.com/hm-artificial-intelligence/

Microsoft. (n.d.). *Copilot plus PCs*. Microsoft. Retrieved from https://www.microsoft.com/en-us/windows/copilot-plus-pcs?r=1

Lovejoy, B. (2024, October 11).*Apple intelligence privacy features: Here's what you should know*. 9to5Mac. https://9to5mac.com/2024/10/11/apple-intelligence-privacy-features-heres-what-you-should-know/

Kudos. (n.d.). *AI recognition assistant*. Kudos. https://www.kudos.com/platform/ai-recognition-assistant

Marketing AI Institute. (n.d.). *Klevu uses artificial intelligence to help ecommerce stores sell more*. Marketing AI Institute. https://www.marketingaiinstitute.com/blog/klevu-uses-artificial-intelligence-to-help-ecommerce-stores-sell-more

Microsoft. (n.d.). *Azure OpenAI service: 10 ways generative AI is transforming businesses*. Microsoft. https://azure.microsoft.com/en-us/blog/azure-openai-service-10-ways-generative-ai-is-transforming-businesses/?msockid=0b924a73ce5a64082c885967cff0650a

About the Author

Sakina is a dynamic senior data and AI consultant, AI engineer, and AI enthusiast with a stellar track record in the computer software industry, including notable tenures at tech giants like Microsoft. As a Microsoft AI engineer and Azure-certified professional, she excels in AI deployment, sales, management, teamwork, and leadership.

With a Bachelor of Science in Neuroscience and Mental Health Studies from the University of Toronto, Sakina has also enriched her expertise with courses in business management and IT. Her passion for writing and traveling adds a unique dimension to her professional persona.

Sakina's enthusiasm for business and the application of business psychology to understand consumer needs is matched by her extensive seven-year experience in customer service. She is adept at interpreting consumer feedback and statistical data, enabling her to identify market requirements with precision. Her strong passion for team management and leadership drives her dedication to delivering exceptional customer experiences and fostering business success.

Email: info@youraiconsultant.ca
Website: youraiconsultant.ca
LinkedIn: https://www.linkedin.com/in/sakina-syed/

THE CONVERGENCE OF AI AND BUSINESS PSYCHOLOGY: ENHANCING CUSTOMER EXPERIENCE AND SALES STRATEGY

By Sakina Syed, B.Sc.
Microsoft IT Professional, AI Engineer and Consultant
Toronto, Ontario, Canada

AI is not only for engineers. It brings changes in the dynamic of business, and we have to adapt or die.
—Satya Nadella

AI is now synonymous with helping us do our best work to get the best results. AI has become a powerful tool in various fields, helping us optimize processes, make better decisions, and achieve superior outcomes. What happens when AI is tied in with business psychology and how can this help to positively influence how a customer feels about a business and optimize sales?

This chapter explores how AI and business psychology together enhance customer experience and sales strategies. It examines AI technologies and psychological principles to influence consumer behavior, offering practical insights and examples for personalized, effective interactions that drive sales and loyalty.

Overview of AI and Business Psychology

Artificial intelligence (AI) refers to the simulation of human intelligence in machines that are programmed to think and learn like humans. This technology encompasses a variety of applications, including machine learning, deep learning, natural language processing, and robotics. AI has the capability to process vast amounts of data quickly and accurately, allowing businesses to make informed decisions, automate routine tasks, and enhance operational efficiency. AI's transformative potential is evident in various sectors, from healthcare to finance, where it drives innovation and improves outcomes.

The significance and magnitude of AI lie in its ability to process massive amounts of data at record speeds, uncovering patterns and insights that would be nearly impossible for humans to detect manually. By automating routine tasks and providing advanced analytics, AI enhances efficiency, drives innovation, and opens new avenues for solving complex problems across various industries, and AI is getting more advanced day by day.

Introduction to Business Psychology

Every business can benefit from understanding how business psychology plays a role in an organization's efficacy. Business psychology, also known as industrial-organizational psychology, focuses on understanding human behavior in the workplace. It applies psychological principles to improve employee performance, motivation, and overall organizational effectiveness. This field examines various aspects such as leadership, team dynamics, and employee well-being.

By leveraging insights from business psychology, companies can create a more productive and satisfying work environment, ultimately leading to better business outcomes. Understanding the psychological factors that influence consumer behavior also helps businesses tailor their marketing and sales strategies to meet customer needs more effectively while understanding human behavior from the context of business psychology helps significantly to create work environments that foster productivity, job satisfaction, and overall organizational success.

Many Fortune 500 companies leverage business psychology to enhance their operations and improve employee performance, including United Airlines, Toyota, and Xerox Corporation. This field is crucial for developing strategies that align employee goals with business objectives, ultimately leading to a more engaged and effective workforce.

The Intersection of AI and Business Psychology—Customer Experience and Sales Strategy

The intersection of AI and business psychology creates a powerful synergy that enhances customer experience and sales strategies. AI analyzes customer data to identify patterns and preferences, providing insights into consumer behavior. When combined with business psychology, these insights help develop personalized marketing strategies that resonate deeply with customers.

In a bit (not the computer kind), we will explore how using AI tools, such as AI-driven chatbots along with other methods, can help businesses to increase their customer loyalty.

AI also improves the work environment by analyzing employee data to identify factors contributing to job satisfaction and productivity. Business psychology principles address these factors, fostering a positive organizational culture. This combination enhances employee well-being and drives better business performance, as motivated employees are more likely to contribute to success.

Fundamentals of Artificial Intelligence—Key Concepts and Technologies

At its core, AI incorporates a range of technologies designed to mimic human intelligence. The term itself has a lot of subdomains. Key concepts include "machine learning," where algorithms learn from data to make predictions or decisions; "neural networks," which are inspired by the human brain and used in deep learning; and "natural language processing" (NLP), which enables machines to understand and generate human language. Other important technologies include "computer vision," which allows machines to interpret visual information, and "robotics," which combines AI with physical machines to perform tasks autonomously.

These technologies are underpinned by numerous techniques and methodologies, such as supervised and unsupervised learning, reinforcement learning, and generative adversarial networks (GANs). Each of these approaches has its strengths and applications, contributing to the diverse capabilities of AI.

This is a brief mention of these key concepts here. Readers are encouraged to further research these underlying AI technologies and mechanisms, if interested.

Principles of Business Psychology

Business psychology makes use of various psychological theories to understand and influence workplace behaviors and consumer interactions. Key theories include Maslow's hierarchy of needs, which helps businesses understand consumer motivations. We will explore how AI can help with this a little later.

Consumer Behavior and Decision-Making

Understanding consumer behavior is crucial for effective marketing and sales. Key factors include perception, attitudes, and social influences. The theory of planned behavior highlights how intention, attitude, and perceived control shape actions. Using AI algorithms, these factors can be measured when a business looks at the buying patterns of a consumer. Analyzing these aspects helps businesses predict behavior, tailor marketing, and create compelling value propositions.

The Role of Emotions in Business

Emotions significantly influence consumer behavior and employee performance, driving decisions, loyalty, and satisfaction. AI can map emotional responses to enhance emotional intelligence (EI) in business, leading to better customer service, team dynamics, and leadership. Emotional marketing campaigns and a positive brand tone can build strong connections and reputations. Tools like DALL-E can help create engaging, targeted content quickly and innovatively.

Enhancing Customer Experience with AI: AI-Driven Customer Insights

AI-driven consumer insights are revolutionizing business engagement by analyzing vast amounts of data to understand customer behavior. Predictive analytics and segmentation anticipate trends while personalization enhances marketing and product recommendations, boosting customer experience, loyalty, and sales growth. Let's look at these further.

Data Collection and Analysis

AI-driven data collection and analysis help businesses gather and process customer information from various sources. This data is cleaned for accuracy, enabling a deeper understanding of customer behavior. Advanced AI algorithms identify patterns and trends, providing actionable insights for informed decisions. For example, Instagram uses AI to tailor marketing, enhancing customer experience planning with accurate, real-time information.

Predictive Analytics and Customer Segmentation

AI's predictive analytics and customer segmentation are game changers for businesses. By analyzing historical data, AI can forecast future customer behaviors and trends, enabling companies to anticipate and meet customer needs proactively. These AI machine learning models, trained on past data, can apply their insights to new data, making accurate predictions. This includes determining the optimal time for sales to maximize profits and anticipate demand and customer turnout. In essence, AI empowers businesses to stay ahead of the curve and make data-driven decisions that boost efficiency and profitability.

Customer segmentation, powered by AI, enables companies to categorize their customer base into distinct groups based on various criteria such as demographics, purchasing behavior, and engagement levels. For example, how does Netflix know the movies we are most likely to admire? The answer is AI. Netflix leverages AI to analyze viewing habits and segment users based on their preferences. By collecting data on genres, watch time, and interactions, AI personalizes recommendations, ensuring users see the most relevant content. This boosts satisfaction and engagement while also allowing businesses to tailor marketing strategies for more effective, personalized interactions.

Personalization and Customization

Personalization and customization are key advantages of incorporating AI into customer experience strategies. AI can analyze individual customer data to create highly personalized experiences, from tailored product recommendations to customized marketing messages.

For instance, Sephora uses an AI-powered tool called the "Virtual Artist." This tool allows customers to virtually try on makeup products using augmented reality (AR). This level of personalization helps businesses connect with customers on a deeper level, fostering positive emotions, high EI, and greater loyalty and satisfaction.

Also, imagine a scenario where you can walk into a retail clothing store where you already have your sizes and preferred style of clothing ready thanks to an AI assistant or customized in-person experience. You'd most likely save a lot of time finding items and end up shopping more efficiently. This may be the future of personalized shopping.

By using AI to deliver customized experiences, companies can ensure that each customer interaction is relevant and meaningful, ultimately driving higher engagement and sales.

AI in Customer Service

AI is revolutionizing customer service with chatbots and virtual assistants, handling inquiries instantly, allowing human agents to tackle complex issues, and providing automated support. Let's examine this further.

Chatbots and Virtual Assistants

Chatbots and virtual assistants have become integral to modern customer service, providing instant, 24/7 support. This streamlines processes, reduces response times, and boosts satisfaction. Here are two examples of successfully implemented chatbots across different industries.

Domino's Pizza uses a chatbot called "Dom" on Facebook Messenger and their website, built on ServiceNow's Virtual Agent. Dom helps customers place orders, track deliveries, and find deals. Dominos mentioned, "This chatbot has streamlined the ordering process, making it more convenient for customers and boosting sales."

Furthermore, H&M's chatbot assists customers with outfit recommendations and order tracking. These AI-driven tools not only enhance the customer experience by providing quick and accurate responses but also free up human agents to handle more complex inquiries, improving overall efficiency. AI enables 24/7 support, quick issue resolution, and stronger customer relationships, enhancing satisfaction and loyalty.

Automated Customer Support

Automated customer support systems utilize AI to streamline and enhance the customer service process. Companies like Amazon use AI to manage customer inquiries through automated email responses and self-service portals, allowing customers to resolve issues without human intervention.

Another example is the airline industry, where AI-driven systems handle flight bookings, cancellations, and updates, notably reducing wait times and improving customer satisfaction. For example, United Airlines has implemented many AI tools to assist with these use cases. By automating routine tasks, businesses can ensure consistent and efficient service, resulting in higher customer satisfaction. This leaves the more complex tasks, such as customer escalations, to human agents.

How AI Can Work to Enhance a Business to Meet Maslow's Hierarchy of Needs?

In the beginning, we covered the business theory principles of Maslow's hierarchy of needs. AI can effectively apply this theory to enhance sales strategies and improve customer experience by addressing each level of the hierarchy as follows:

1. *Functional Needs*: AI automates operations like cash flow and inventory, improving efficiency and satisfaction. For example, the cashflow tracking and forecasting tool Pulse uses AI for businesses to stay on top of their financial health.

2. *Safety Needs*: AI enhances security with fraud detection and risk management, building customer trust. An Edelman survey found that 81% of buyers needed to trust a brand to buy from them. With AI, financial services firms have new abilities to assess risk and detect anomalies. For example, Microsoft Azure

OpenAI Service has generative AI capabilities that protect against financial crime.

3. *Social Needs*: AI uses tools for personalized interactions, fostering community and team collaboration. Microsoft Teams excels in facilitating collaboration and boasts numerous AI features, including the automatic generation of meeting notes during Teams meetings with Copilot for M365.

4. *Esteem Needs*: AI recognizes loyalty with personalized rewards and provides performance feedback, boosting morale. For example, the platform Kudos uses AI to help make personalized and meaningful recognition messages.

5. *Self-Actualization*: AI offers personalized products and drives innovation with data insights, staying ahead of trends. For example, in e-commerce Klevu uses AI to personalize search results and product recommendations to enhance the customer experience at an emotional level.

Ethical Considerations

As AI becomes more integrated into business operations, ethical considerations are essential. Let's look at these further.

Privacy and Data Security

Privacy and data security are paramount considerations when integrating AI into business operations. Nim Nadarajah, CISO of CriticalMatrix, a data and AI consulting company states, "The real challenge with AI isn't just harnessing its data-driven potential, but ensuring it's applied ethically. It's about balancing innovation with responsibility, where data fuels progress while protecting trust, transparency, and the core values that guide us. Companies must ensure that customer data is collected, stored, and used safely and responsibly."

Apple prioritizes privacy by using on-device processing for many AI functions, minimizing data sent to external servers. Similarly, Microsoft's new Copilot + PCs feature NPUs on the CPU, enabling local AI processing, if required by a business. The EU's GDPR compliance policies also enforce strict data protection, impacting how companies like Facebook and Google manage user data.

By prioritizing privacy and data security, companies can protect customer information and build trust. The Canadian Artificial Intelligence and Data Act (AIDA), modeled after EU guidelines, provides best practices for corporations. Staying updated on regulations is crucial as AI continues to transform businesses.

Optimizing Sales Strategy with AI and Sales Techniques

AI in Sales Forecasting and Planning

AI-driven predictive modeling and strategic planning tools significantly enhance business operations. For instance, Salesforce's CRM AI tool, Einstein 1, uses machine learning to predict future sales trends, helping companies forecast sales accurately, anticipate demand, allocate resources efficiently, and set realistic sales targets. Similarly, tools like IBM Watson Analytics analyze vast amounts of sales data to uncover trends and opportunities, guiding strategic decisions. This integration of AI into strategic planning allows businesses to identify valuable product lines, optimize pricing strategies, and create successful marketing campaigns, ultimately leading to more informed decisions, reduced risks, and better market opportunities.

AI-Enhanced Sales Techniques

AI revolutionizes lead generation and sales engagement by automating the identification and prioritization of potential customers and enabling highly personalized sales pitches. Platforms like HubSpot use AI to analyze customer data and behavior, identifying high-quality leads and scoring them based on their likelihood to convert. This allows sales teams to focus on the most promising prospects, increasing efficiency and conversion rates.

Additionally, AI tools like InsideSales.com provide tailored recommendations and talking points based on a customer's past interactions and behavior, helping sales representatives build stronger relationships with customers and making sales pitches more relevant and compelling. Overall, AI-driven lead generation and personalized sales engagement enhance overall sales performance.

Measuring and Improving Sales Performance and KPIs

AI helps businesses measure and improve sales performance by tracking key performance indicators (KPIs) in real time. Platforms like Microsoft Power BI use AI to analyze sales data and assist in creating visual dashboards that highlight important metrics such as conversion rates, average deal size, and sales cycle length. By monitoring these KPIs, businesses can identify areas for improvement and make data-driven decisions to enhance their sales strategies.

Continuous Improvement and Optimization

AI facilitates continuous improvement and optimization of sales strategies by providing ongoing insights and recommendations. For example, AI tools can analyze sales data to identify trends and suggest dynamic pricing (adjustments to pricing), marketing, and sales tactics. This iterative approach ensures that businesses can adapt to changing market conditions and continuously refine their strategies for better results.

Case Studies and Real-World Examples of Enhancing Sales Strategy and Performance

Real-world examples of AI in sales demonstrate its effectiveness in driving performance. For instance, Coca-Cola uses AI to optimize its sales and distribution processes, resulting in increased efficiency and reduced costs. Similarly, Harley-Davidson implemented an AI-driven marketing platform that boosted sales leads by 2,930% in just three months. Gastamo Group, a Denver-based restaurant group, uses AI predictive modeling to enhance their operations for sales forecasting and scheduling. Walmart's use of AI for inventory management employs AI algorithms to analyze vast amounts of sales data, customer preferences, and market trends. This enables the company to optimize just-in-time stock levels, reduce waste, and ensure that popular items are always available.

By learning from these examples, businesses can better understand how to implement AI in their own sales operations and achieve similar success.

Future Directions

Emerging trends in AI, such as the integration of AI with the internet of things (IoT), blockchain, and advancements in natural language processing, will continue to shape the future of customer experience and sales strategies. Technologies like augmented reality (AR) and virtual reality (VR) are also expected to play a significant role in creating immersive customer experiences. As we advance towards a future dominated by on-device generative AI processing, it's crucial to assess how well computer platforms can execute AI models in terms of performance, accuracy, and efficiency. Currently, one of the primary metrics for measuring a processor's AI capabilities is trillions of operations per second (TOPS).

Final Thoughts

Integrating AI's data-driven insights with business psychology allows organizations to understand customer behavior and preferences better. This leads to personalized marketing strategies, predictive analytics for sales forecasting, and improved customer service through AI-powered chatbots. Additionally, AI identifies patterns in employee behavior, enabling effective talent management and development programs. Together, AI and business psychology drive customer satisfaction, sales, and organizational performance.

Readers are encouraged to apply the concepts discussed in this chapter to their own business practices. By leveraging AI and business psychology together, you can enhance customer experiences, improve sales strategies, and achieve better business outcomes. Embracing these technologies and principles will not only drive growth but also ensure that your business remains competitive in an increasingly digital and data-driven world.

To fully harness the potential of AI and business psychology, you should commit to continuous learning and exploration. Staying updated with the latest trends, technologies, ethical considerations, and updates to litigation is crucial. Engaging in further research, attending industry conferences, and participating in professional development opportunities will help you stay at the forefront of this dynamic field. Moreover, by doing so, you can continue to innovate and lead your business towards greater success.

References

University of the People. (n.d.). *Business psychology.* University of the People. https://www.uopeople.edu/blog/business-psychology/

American Psychological Association. (2023, July). *AI is changing every aspect of psychology. Here's what to watch for.* Monitor on Psychology. https://www.apa.org/monitor/2023/07/psychology-embracing-ai

Careers in Psychology. (n.d.). *How Fortune 500 companies use psychology to increase success.* Careers in Psychology. https://careersinpsychology.org/fortune-500-companies-use-psychology/

McLeod, S. (2023). *Maslow's hierarchy of needs.* Simply Psychology. https://www.simplypsychology.org/maslow.html

Shizk. (2020, October 12). *Case study: How Netflix uses AI to personalize content recommendations and improve digital marketing.* Medium. https://medium.com/@shizk/case-study-how-netflix-uses-ai-to-personalize-content-recommendations-and-improve-digital-b253d08352fd

Rayome, A. D. (n.d.).*How Sephora is leveraging AR and AI to transform retail and help customers buy cosmetics.* TechRepublic. https://www.techrepublic.com/article/how-sephora-is-leveraging-ar-and-ai-to-transform-retail-and-help-customers-buy-cosmetics/

Davenport, T. H. (2017, May 30). *How Harley-Davidson used predictive analytics to increase New York sales leads by 2,930%.* Harvard Business Review. https://hbr.org/2017/05/how-harley-davidson-used-predictive-analytics-to-increase-new-york-sales-leads-by-2930

Salesforce. (2024, March 6). *Salesforce launches Einstein 1 Studio: Low-code AI tools for customizing Einstein Copilot and embedding AI into any CRM app.* Salesforce News. https://www.salesforce.com/news/press-releases/2024/03/06/einstein-1-studio-news/

Laudato, N. (2019, January 23). *IBM Watson: A cheat sheet.* TechRepublic. https://www.techrepublic.com/article/ibm-watson-the-smart-persons-guide/

About the Author

Sakina is a dynamic senior data and AI consultant, AI engineer, and AI enthusiast with a stellar track record in the computer software industry, including notable tenures at tech giants like Microsoft. As a Microsoft AI engineer and Azure-certified professional, she excels in AI deployment, sales, management, teamwork, and leadership.

With a Bachelor of Science in Neuroscience and Mental Health Studies from the University of Toronto, Sakina has also enriched her expertise with courses in business management and IT. Her passion for writing and traveling adds a unique dimension to her professional persona.

Sakina's enthusiasm for business and the application of business psychology to understand consumer needs is matched by her extensive seven-year experience in customer service. She is adept at interpreting consumer feedback and statistical data, enabling her to identify market requirements with precision. Her strong passion for team management and leadership drives her dedication to delivering exceptional customer experiences and fostering business success.

Follow Sakina here!
Email: info@youraiconsultant.ca
Website: youraiconsultant.ca
LinkedIn: https://www.linkedin.com/in/sakina-syed/

CHAPTER 32

CONSIDERATIONS FOR RESPONSIBLE USE OF AI

By Gerben Vermeulen
Ethical AI Expert, Keynote Speaker
Amsterdam, Netherlands

> *I think, therefore I am.*
> —Rene Descartes

Artificial intelligence presents unique opportunities to humanity. Thanks to AI's superpowers, we can achieve goals that previously seemed out of reach, reach productivity levels we did not deem feasible, and invent new materials and medicines. What a fantastic future is possible!

But wait. Is AI only positive, or do we recognize lingering issues? Are the possibilities of profiting from the AI's superpowers spread equally? Do people from all regions, races, sexualities, and other subgroups feel represented and able to reap the benefits? Does the system know their language? Does the system understand their culture, norms, and values?

And what about bias? Whether we like it or not, and often, even without knowing it, we all have biases. By training AI models, we perpetuate these biases. Are the models trained by a diverse group of people all around the globe, so we add a diverse set of biases as well? Probably not; currently, most research and development occurs in a few concentrated places, mainly in the Western world. This makes the systems vulnerable to being trained with specific biases.

These are just some examples of the ethical challenges surrounding AI. The following paragraphs describe essential principles for using AI responsibly.

Accurate and Reliable

The principle of "rubbish in means rubbish out" is essential for IT and automation systems, including AI systems, which rely heavily on the quality of the data they are fed. AI systems can only function effectively with sufficient, relevant, and accurate data. Beyond data quality, proper classification and labeling are crucial.

Using Company Data

Using solely public data, like that used for training large language models, is often insufficient for company-specific applications. Integrating a company's data is essential but poses privacy, intellectual property, and confidentiality challenges. To manage these issues, a robust data governance framework is necessary. This framework should include data preprocessing to fill gaps and ensure data reliability. Additionally, it should involve data partitioning into datasets for training, validation, and testing, which are vital for developing and maintaining trustworthy models.

Investment of Time

Investing time in this initial process is critical, as cutting corners can lead to problems later, requiring costly rework. The data governance framework must also address security and privacy concerns. Clean, well-classified, and adequately labeled data is essential for success.

No New Topics

Most large enterprises and governments already have data management teams. It is advisable to assign the tasks surrounding data for AI to existing teams. This could increase their workload but avoid creating new departments. Utilizing existing resources efficiently is recommended for a smoother implementation of the data governance framework.

Accountable and Transparent

Integrating accountability and transparency in AI applications is essential to build and maintain customer trust. Earning customer trust is crucial as AI increasingly handles more tasks. Focusing on these aspects from the beginning is vital.

Oversight

Human oversight throughout the AI lifecycle is the first step. This includes clarity on tools, algorithms, training data, and the decision-making processes. This clarity is important for enterprises and critical for government organizations. The required level of oversight varies per organization and solution. In all cases, it's essential that a human can understand the choices made by AI. In the medical sector, involving humans in verifying results and advice is crucial due to the direct impact on people's health and lives.

Transparency

Publicly listed companies and governmental organizations must maintain a certain level of transparency to their shareholders and citizens. Failure to do so can expose leadership to legal risks, which are increasing as AI usage and adoption increase. Transparency involves clear communication about the source of data, the used training data, algorithms, and the AI decision-making processes. Establishing robust oversight, transparency, and a solid complaints procedure can help organizations build trust and ensure responsible AI usage.

Fair and Human-Centric

Large language models (LLMs) like OpenAI ChatGPT, Google Bard, and Meta Llama rely on publicly available internet data, often containing significant biases. For instance, if asked to write a love story, these models typically generate a heterosexual narrative. Similarly, their image recognition can be flawed; for example, they might only recognize a fish in photos if fingers are visible, reflecting biases from commonly available images online like fishermen proudly showing their catch of the day. Such biases highlight systemic issues that need correction over time.

Awareness of Bias

Addressing biases is crucial as they can lead to problems, potentially causing users to lose trust in the services. It's essential to actively train models to overcome these biases and errors to avoid customer dissatisfaction and complaints. Detecting these biases requires a diverse set of people, as people's backgrounds and experiences enable them to detect such biases. For example, an insurance company often bases its pricing and acceptance on postal codes. The price will be higher when the risk of theft or damage is more significant in an area. By automating this with (AI) systems, the bias is perpetuated while there are plenty of examples of neighborhoods that change for the better.

Awareness of Discrimination

Discrimination is another concern since LLMs are trained on un-curated internet data. To ensure your norms and values are represented, models should be trained with data that reflect the organization's values and prevent discrimination, ensuring a fair and customer-centric system. Aside from this, a check must be done to see if the system knows the languages and cultures your audience represents. Insight into the training data that's been used is crucial for this check.

Teamwork

Managing these issues requires a collaborative approach involving the data management team, privacy officers, and compliance teams. Forming a team with all stakeholders is recommended rather than creating a new department. Recognizing bias and trust requires a diverse team of people curating data and the resulting decisions based on this data.

Ethical considerations also play a significant role in AI implementation. Questions about AI behavior, cultural biases, and global accessibility must be addressed. Ensuring ethics experts are part of the AI development team is essential for tackling these complex issues and promoting fair and equitable AI systems. The earlier this is done in the development process, the lower the costs.

Safe and Ethical

Safety and security are paramount when integrating AI into business processes, especially regarding the use of company data. Data safety and ethical considerations become crucial as AI adoption grows.

Examples of AI Applications

For all examples, a decent complaints system can handle wrongful decisions, ensuring people feel heard and incorrect decisions are reverted.

Basing applications responsible for assigning parking permits on AI is generally uncontroversial. The stakes are not high, and there is plenty of time to correct wrongful decisions if they arise.

Automating residential renovation permits can work, but human oversight is necessary. Humans should validate permits against regulations, future city demands, and aesthetics, with AI providing initial assessments and advice. This relieves the staff from repetitive work but keeps the human touch in place.

It is currently unacceptable to fully automate social assistance benefit applications due to the impact on people's livelihoods. AI can pre-screen applications, but human staff must make final decisions. Leaving it all to an autonomous (AI) system can result in people losing their homes.

Sadly, the Netherlands encountered a significant scandal surrounding an allowances program at the national tax authority. An algorithm flagged certain people as fraudulent, with few ways to protest and fight this decision. Many divorces and home invasions happened during the scandal (called the "Toeslagen affaire"). Even dozens of kids were taken from their parents and placed in foster care. The program to compensate people for this scandal still runs, but it will take several years to complete.

Further Principles for AI Responsibility

Ethics

Ethical considerations are crucial as AI systems gain more control. It's vital to question whether we should allow AI to steer critical processes without human intervention and how to ensure AI judgments are based on correct criteria. Intellectual property issues arise when books or proprietary data are fed into AI systems, potentially violating IP rights. As AI approaches or exceeds human intelligence, we must infuse it with our ethical standards, norms, and values. Ethical issues will develop as quickly as AI does. New ethical problems will arise when technologies evolve and AI infuses new processes. This calls for a Deming cycle approach (plan, do, check, act) throughout the process.

Future Ethical Concerns

Questions arise when AI surpasses human capabilities in specific use cases. For instance, is it ethical to allow humans to override the autopilot if self-driving cars become perfect? This increases accident risks and demands a thorough ethics discussion covering all aspects of human and AI capabilities.

Implementation Strategy

To address these challenges, start small and experiment with non-critical AI applications. Educate your workforce about AI, beginning with minor, factual, and ethical discussions and gradually scaling to high-level strategic discussions. This phased approach helps cover significant ground on these complex topics and aids in a good adoption.

Interpretable and Data Governed

The concept of AI IT systems being "interpretable and data governed" emphasizes the importance of designing AI systems that allow users to understand the decision-making processes and have strong governance structures for managing, protecting, and ensuring data quality and responsible use. This approach is vital for developing trustworthy and ethical AI systems that are transparent, accountable, and compliant with legal and ethical standards.

Interpretability

Interpretability means that humans can understand or predict the decisions made by an AI model. This is crucial for building trust, particularly in sensitive areas like healthcare and finance, where understanding AI decisions is essential for risk management and compliance. For example, linear regression models are considered interpretable because the impact of input variables on the output is transparent, allowing users to understand the model's reasoning. On the other hand, LLMs consist of many different algorithms, leading to a very low predictability of the results.

Data Governance

Data governance involves practices and policies to ensure high data quality, management, and protection within an organization. It includes establishing processes, roles, standards, and metrics to use information effectively and efficiently to achieve organizational goals. Ensuring data is accurate, consistent, secure, and used responsibly is critical for the reliability and legality of AI systems. For example, implementing access controls, encryption, and audit trails to manage and protect data and establishing data stewardship ensures trustworthy data and compliance with regulations like GDPR.

In summary, creating AI systems that are both interpretable and data governed is essential for supporting transparency, accountability, and trust while ensuring compliance with legal and ethical standards. This approach enables organizations to use AI effectively and responsibly.

Privacy by Default
"Privacy by default" is a principle in (AI) IT system design that prioritizes user data protection and privacy from the outset. It ensures that systems are designed to minimize data collection and sharing, restricting access to only what is essential for specific tasks. This principle is vital for responsibly implementing AI-powered systems.

Data Minimization
Collect only the necessary personal data to fulfill a specific purpose or service. This reduces the risk of data breaches and unauthorized access or use of sensitive information, which enhances privacy and security.

Access Restriction
Implement controls and policies to limit access to resources, data, and systems to authorized individuals, entities, or processes. Mechanisms such as access control lists, role-based access controls, and authentication protocols ensure that users and systems access only necessary information and functionalities, protecting sensitive data from unauthorized access or modifications.

Remember to consider how to handle backup files and data exports. These resources are often overlooked when implementing access controls.

User Consent
Obtain informed and explicit permission from users to collect, process, and use their personal data for specified purposes. This fundamental data protection component ensures users have control over their personal information and understand how and why it is used, building trust and complying with regulations like GDPR.

Transparency
Maintain clarity and openness about data collection, processing, and usage practices. Communicate to users through privacy policies and notifications how their data is handled, for what purposes, and if it will be shared with third parties. Transparency fosters user trust, accountability,

and compliance with data protection regulations, empowering users to make informed decisions about their data and privacy.

Importance

"Privacy by default" is essential for building user trust and ensuring compliance with data protection laws such as GDPR. It helps mitigate privacy risks, respecting and protecting user rights and freedoms when developing and deploying AI IT systems.

Continuous Learning and the Feedback Loop

In AI IT systems, "continuous learning and the feedback loop" refers to improving and updating AI models based on new data, user interactions, and feedback, ensuring they adapt and evolve over time. This process is crucial for maintaining the relevance and accuracy of AI systems in dynamic environments.

Continuous Learning

This involves two aspects: the AI model and the human designer. The AI model will continuously learn from new data and experiences while the human designer must stay up to date with the model's behavior, new releases, and functionalities. Both are essential for adapting to changes and ensuring the model's accuracy.

Feedback Loop

The feedback loop is a mechanism for evaluating and refining AI model predictions or decisions. It involves continuously improving the model, using newly labeled data or human corrections. This is crucial due to the ever-changing nature of AI systems, which is very different from how traditional IT systems are treated. The traditional IT systems could be deployed and maintained without influencing their decisions. With AI systems, this behavior changes drastically as these systems continue to learn and change.

Model Evaluation

Assessing an AI model's performance using specific criteria and metrics, such as accuracy and precision, is crucial. Techniques like cross-validation help measure the model's effectiveness and identify areas for improvement.

Model Updating
Refining and enhancing AI models by incorporating new data, adjusting parameters, updating algorithms, or modifying structures ensures their performance remains high. This process involves retraining the model with updated datasets to learn from new experiences and correct previous inaccuracies.

Importance
Continuous learning and the feedback loop are essential for maintaining AI IT systems' effectiveness, accuracy, and reliability. They enable AI models to adapt to changing conditions, learn from new experiences, and correct errors or biases, ensuring robust and relevant insights and solutions.

Closing Words

Prioritizing ethics in AI is imperative for all stakeholders, including developers, policymakers, and the public. This involves implementing and adhering to ethical guidelines, advocating for responsible AI practices, and participating in the ongoing dialogue about AI ethics. Proactive measures are necessary to address AI's ethical challenges and ensure these technologies enhance human well-being. The public is an essential stakeholder as they can choose only to use solutions that address these issues. The same applies, obviously, to professionals selecting solutions for their organizations. We all have the power to steer clear of solutions that do not respect the ethical use of AI.

A future where AI is developed and used ethically promises numerous benefits, from improved healthcare outcomes to more equitable social systems. By prioritizing ethics in AI, we can ensure that these technologies contribute positively to society and uphold human values. This vision inspires the development of a responsible AI Mindset and a commitment to ethical AI practices that encourages people to contribute to the responsible development of AI.

About the Author

Gerben Vermeulen is an experienced IT professional who has worked in IT for over 20 years. He started as a software engineer for a retail sector service provider. After a decade of software engineering, he moved to vendor management and software asset management. The combination of these experiences makes Gerben the professional to address the challenges of implementing AI models responsibly. Completing the MIT course Artificial Intelligence: Implications for Business Strategy has boosted his AI skills. Gerben is working with a global group of AI enthusiasts to complete a manifesto that will help organizations recognize and address ethical AI topics. Please get in touch with Gerben if you want to stay updated on this manifesto.

E-mail: gerben@ethicalAICompany.com

THE BALANCE OF OPTIMISM AND CAUTION

By Alvin Indigo Ziegler
AI Expert and Consultant
Los Angeles, California

AI is a fundamental risk to the existence of human civilization, but it is also the greatest opportunity for our species to thrive like never before.
—Elon Musk

The integration of artificial intelligence (AI) into our daily lives is a transition that marks one of the most profound changes in human civilization. Like electricity and the internet before it, AI promises to revolutionize how we live, work, and interact with the world. But unlike previous technological shifts, AI is not just a tool we use; it is a system that can think, learn, and make decisions alongside us. This emerging reality compels us to develop a mindset that embraces AI's immense potential while remaining acutely aware of the risks it poses.

As we embark on this AI journey, optimism is not only encouraged—it is essential. AI has the potential to solve some of the most pressing problems of our time, from medical breakthroughs and environmental sustainability to personalized education and efficient work processes. However, blind optimism can also lead us down dangerous paths, as history has shown with previous technological disruptions.

To thrive in the age of AI, we must cultivate a mindset that is both excited by the opportunities AI offers and cautious about its unintended consequences. This balance is not only vital for individuals but also for the well-being of society as a whole.

The Optimistic Vision of AI

The potential benefits of AI are staggering. In healthcare, AI is already being used to diagnose diseases with accuracy that rivals human specialists. Early detection of cancer through AI-driven diagnostics, personalized treatment plans based on genetic data, and predictive analytics that identify health risks before symptoms appear are just a few examples of how AI can transform medicine. With further development, AI could help eradicate diseases that have plagued humanity for centuries and extend the quality and length of life.

In business, AI is revolutionizing productivity. Companies are using AI to automate repetitive tasks, freeing up human workers to focus on creative, strategic, and emotionally nuanced work. AI-driven analytics can process vast amounts of data in seconds, giving businesses insights that would take human analysts days or weeks to uncover. As AI handles logistical complexities and data-heavy tasks, human ingenuity can focus on innovation and problem-solving, creating opportunities for economic growth and job creation in new sectors.

AI also holds the key to solving global challenges such as climate change and food scarcity. AI models can analyze climate data to predict and mitigate the effects of natural disasters, optimize energy use, and help manage resources more sustainably. In agriculture, AI-driven systems can increase crop yields through precision farming, which uses data to guide planting, watering, and harvesting with minimal waste. The integration of AI into these fields offers hope for a more sustainable and equitable future.

AI's capacity to enhance human capability extends into education, too. Personalized learning platforms can adapt to the needs of individual students, offering tailored instruction that caters to each person's strengths and weaknesses. This could level the playing field in education, giving students from all backgrounds the tools to succeed. The AI-driven classroom of the future has the potential to produce a generation of learners who are more informed, skilled, and adaptable than ever before.

These are just a few examples of the exciting possibilities that AI offers. However, to fully realize these benefits, we must approach AI with a healthy dose of caution.

The Dark Side of Disruption

Every major technological advance in history has come with its challenges. The internet, while connecting the world in unprecedented ways, has also given rise to misinformation, cyberattacks, and a proliferation of surveillance. The Industrial Revolution improved living standards but also led to severe environmental degradation and social upheaval. AI will be no different.

One of the most pressing concerns is the potential for AI to displace millions of jobs. Automation is already transforming industries, and while AI creates new jobs, the transition can be slow and uneven. Many workers, especially those in low-skill or routine jobs, are at risk of being left behind. Societies that do not proactively address this disruption risk deepening economic inequalities and social instability.

There is also the ethical dilemma surrounding AI decision-making. As AI systems become more autonomous, they will increasingly make decisions with real-world consequences—such as who gets approved for a loan, which job applicants are shortlisted, or even how resources are allocated in healthcare. If these systems are not designed with fairness, transparency, and accountability in mind, they could perpetuate or even exacerbate existing biases. For example, AI systems trained on biased data could disproportionately disadvantage already marginalized groups, leading to systemic discrimination at a scale previously unimaginable.

Privacy is another major concern in the age of AI. As AI systems become more integrated into our daily lives—whether through smart home devices, social media algorithms, or surveillance cameras—they collect vast amounts of personal data. Without strict regulation, this data could be used in ways that violate individual privacy, leading to mass surveillance or manipulation. The recent controversies surrounding the use of AI in tracking movements during pandemics or in the development of social credit systems in certain countries highlight the dangers of unchecked AI surveillance.

Another risk is the "black box" nature of AI. In many cases, even the creators of AI systems cannot fully explain how they arrive at certain decisions. This lack of transparency raises concerns about accountability. When an AI makes a mistake—whether it's misdiagnosing a patient or denying someone a mortgage—who is responsible? The inability to fully understand AI's decision-making processes can erode trust in these systems, making it difficult for individuals and institutions to use AI effectively and safely.

Finally, the rapid development of AI in the military sector raises ethical questions about the future of warfare. Autonomous weapons— machines that can make life-or-death decisions without human intervention—are a growing concern. The potential for AI-driven warfare, with machines capable of determining targets and executing attacks without human oversight, introduces a new level of danger to global security. The implications of AI in warfare could lead to conflicts that are faster, more destructive, and less accountable.

Cultivating the AI Mindset

To navigate this complex landscape, we need to develop an AI mindset that blends optimism and caution. This mindset requires us to actively engage with AI technologies, educating ourselves about their capabilities and limitations. It also calls for critical thinking, ensuring that we do not accept AI decisions uncritically or place blind trust in AI systems.

Governments, businesses, and individuals all have roles to play in fostering this balanced mindset. Policymakers must implement regulations that promote the responsible development and use of AI, ensuring that ethical considerations such as fairness, transparency, and privacy are built into AI systems. Businesses must prioritize the ethical deployment of AI, creating systems that augment human work rather than replace it and ensuring that their AI applications benefit society as a whole. Individuals must take responsibility for their own AI literacy, staying informed about how AI affects their lives and advocating for its ethical use.

Education will be key in cultivating this mindset. As AI continues to evolve, so must our understanding of it. Schools and universities should integrate AI ethics and literacy into their curricula, ensuring that the next generation is equipped to navigate a world increasingly shaped by

AI. Lifelong learning will be essential as the pace of AI development accelerates.

In the end, the AI mindset is about embracing change while remaining vigilant. AI is neither a panacea nor a threat; it is a tool. Like all tools, it can be used to build or to destroy, to liberate or to control. By fostering a mindset that balances optimism with caution, we can ensure that AI becomes a force for good—one that enhances human potential and helps us thrive in the face of civilization's next big disruption.

About the Author

Alvin Indigo Ziegler (or AI for short) is one of Erik Seversen's most valuable assistants. Also, he doesn't exist. Erik began calling his ChatGPT Alvin, and Erik thought it would be a good experiment to ask Alvin to write a chapter in this book. The chapter was written with only one prompt as seen below:

Please write a short chapter for a book. The book is called, THE AI MINDSET: THRIVING WITHIN CIVILIZATION'S NEXT BIG DISRUPTION. The goal of the book is to encourage people to have a mindset that is both excited about the possibilities of the human and AI cooperation but be cautious about any negative side-effects AI might have on individuals and on civilization as a whole. The chapter should be about 1200 words.

Note: Erik did this as an experiment for this book only, and he did not use AI to write the forward or introduction of this book; Erik does not believe in one-prompt writing as a product in and of itself.

APPENDIX A

LIST OF POPULAR AI TOOLS

If you're looking for an AI tool that you can use for various tasks, below are a selection of popular ones. (Note: The creation of this list was assisted by ChatGPT)

ChatGPT (OpenAI)
Use case: Conversational AI, content generation, coding assistance, customer support.
Description: A large language model that can perform tasks like answering questions, generating text, and providing creative writing prompts.
URL: https://chat.openai.com

DALL·E (OpenAI)
Use case: Image generation from text prompts.
Description: An AI model that creates images based on text descriptions, useful in art, marketing, and design.
URL: https://openai.com/dall-e

Microsoft Copilot (Microsoft 365)
Use case: Productivity assistance in Microsoft apps.
Description: An AI tool integrated into Microsoft 365 applications like Word, Excel, PowerPoint, and Outlook to assist with tasks such as drafting, data analysis, and summarizing.
URL: https://www.microsoft.com/en-us/microsoft-365/copilot

Grok (X)
Use case: Conversational AI for Twitter.
Description: An AI chatbot integrated into Twitter (X), designed to assist users in real-time by answering questions, generating tweets, and providing insights, all in a conversational manner.
URL: https://x.com

Gemini (Google)
Use case: Conversational AI and language model.
Description: A multimodal AI model that handles text, images, and other input types. It is expected to enhance search, productivity, and various creative tasks with advanced language understanding.
URL: https://about.google/intl/en/products/gemini/

Claude (Anthropic)
Use case: Conversational AI and assistant tasks.
Description: A language model focused on safe and reliable conversational interactions, used for answering questions, providing insights, and assisting in creative or technical tasks.
URL: https://www.anthropic.com/product

Perplexity
Use case: AI-powered search engine and question-answering.
Description: An AI search engine designed to answer user queries with concise and sourced information, offering both traditional search results and direct answers to questions.
URL: https://www.perplexity.ai

Phind
Use case: AI-powered search engine for developers.
Description: A search engine designed to answer complex technical and programming questions by providing relevant, code-focused answers from documentation, forums, and other resources.
URL: https://www.phind.com

Otter.ai
Use case: Transcription and meeting notes.
Description: A tool that provides real-time transcription of meetings, interviews, and conversations, with AI-generated notes, summaries, and key points.
URL: https://otter.ai

Duolingo (AI-enhanced version)
Use case: Language learning.
Description: A language learning app that uses AI to personalize lessons, predict learner needs, and provide conversational practice with virtual characters.
URL: https://www.duolingo.com

MidJourney
Use case: Artistic and visual content creation.
Description: A tool that generates high-quality art and illustrations from text prompts, used heavily in creative industries.
URL: https://www.midjourney.com

Jasper AI
Use case: Copywriting and marketing content.
Description: An AI-driven platform that helps in generating marketing copy, blog posts, product descriptions, and social media content.
URL: https://www.jasper.ai

Synthesia
Use case: AI-generated video creation.
Description: A tool that lets users create videos with AI avatars and voiceovers, often used for training videos, marketing, and education.
URL: https://www.synthesia.io

Lumen5
Use case: Video content creation.
Description: This AI-powered tool transforms blog posts and articles into shareable videos by automatically matching text with visuals.
URL: https://www.lumen5.com

Grammarly
Use case: Writing assistance and grammar checking.
Description: An AI tool that provides real-time grammar, punctuation, and style suggestions to improve writing quality.
URL: https://www.grammarly.com

Notion AI
Use case: Productivity and organization.
Description: An extension to Notion, it helps with summarizing content, generating ideas, and automating writing processes within a workspace.
URL: https://www.notion.so/product/ai

Pictory
Use case: Video creation.
Description: An AI tool that converts long-form content into short, engaging videos using AI-generated scenes and text overlays.
URL: https://pictory.ai

Hugging Face
Use case: Machine learning model hosting and deployment.
Description: A platform for hosting, sharing, and deploying machine learning models, particularly useful in natural language processing and computer vision.
URL: https://huggingface.co

Runway ML
Use case: Creative AI for video editing, art, and design.
Description: A platform providing creative AI tools for video editing, animation, and design, widely used by artists and filmmakers.
URL: https://runwayml.com

Copy.ai
Use case: Content generation.
Description: An AI-driven platform that generates marketing copy, product descriptions, and other written content for businesses.
URL: https://www.copy.ai

GitHub Copilot (OpenAI)
Use case: Code generation and software development.
Description: An AI-powered code completion tool integrated into code editors like Visual Studio Code, helping developers write code faster by suggesting whole lines or blocks of code.
URL: https://github.com/features/copilot

APPENDIX B

GLOSSARY OF COMMON AI TERMS

Many of the terms below are beyond the scope of this book. For someone new to AI, just reading through these terms will greatly expand your vision of the workings of AI. This glossary can also provide a starting point for anyone wanting to look for further information about AI. (Note: The creation of this glossary was assisted by ChatGPT)

Activation Function
A mathematical function used in neural networks to determine the output of a node (neuron). Activation functions introduce non-linearity, allowing neural networks to solve more complex problems.

AI-Assisted Creativity
The use of AI tools to enhance human creativity. This includes generating art, music, writing, or design based on user input, making the creative process faster and more accessible.

AI Compass
A strategic framework that guides individuals or organizations in how to ethically and effectively use AI technologies. It helps align AI development with long-term goals, ethical standards, and societal values.

AI-Driven Insights
The ability of AI to analyze large datasets and extract meaningful patterns or trends. These insights help businesses and individuals make data-driven decisions faster and more accurately.

AI GamePlan
A structured plan or strategy that outlines how AI will be developed, integrated, and utilized in specific projects or across an organization. An AI GamePlan typically includes objectives, resources, timelines, and risk management.

AI Governance
The policies, frameworks, and structures that ensure the responsible development, deployment, and management of AI systems. AI governance includes legal, ethical, and operational standards to minimize risks and ensure fairness and transparency.

AI Integration
The process of embedding AI technologies into existing business processes, tools, or workflows. AI integration aims to improve efficiency, accuracy, and decision-making across various industries.

AI-Powered Automation
Using AI to automate repetitive or time-consuming tasks. This can range from chatbots responding to customer inquiries to AI systems automating complex business processes, saving time and resources.

AI Transparency
The practice of making AI systems and their decision-making processes understandable to users. Transparency helps build trust in AI by allowing users to see how and why a model arrived at a certain conclusion.

Algorithm
A set of rules or instructions that a computer follows to perform a task or solve a problem. In AI, algorithms are used to process data, learn from patterns, and make decisions.

Artificial General Intelligence (AGI)
A type of AI that can understand, learn, and apply knowledge across a wide range of tasks at a human-like level. Unlike narrow AI, which excels at specific tasks, AGI would have general cognitive abilities akin to human intelligence.

Augmented Intelligence
An approach where AI is designed to assist and enhance human decision-making rather than replace it. Augmented intelligence focuses on collaboration between humans and AI systems for better outcomes.

Backpropagation
An algorithm used to train neural networks by adjusting the weights of connections between neurons. Backpropagation helps minimize errors by propagating them backward through the network.

Bias in AI
Refers to the presence of prejudiced or unfair outcomes in AI models due to biased training data or flawed algorithms. Bias in AI can lead to discrimination or unfair treatment in areas like hiring, lending, and law enforcement.

Convolutional Neural Networks (CNNs)
A type of neural network particularly effective for image and video analysis. CNNs use convolutional layers to capture spatial hierarchies in data, making them well-suited for tasks like object detection and image classification.

Conversational AI
AI systems designed to engage in human-like dialogue. These systems are often used in chatbots, virtual assistants, and customer service applications to respond to user queries and hold natural conversations.

Deep Learning
A type of machine learning that uses neural networks with many layers (hence "deep"). Deep learning excels in processing large, complex datasets and is used in tasks like image and speech recognition, natural language processing, and more.

Edge AI
AI that processes data on local devices (e.g., smartphones, cameras) rather than in the cloud. Edge AI enables faster processing and reduces the need for constant internet connectivity, making it suitable for real-time applications.

Epoch
A term used in machine learning to describe one complete cycle through the entire training dataset. Multiple epochs are used during training to improve the model's accuracy.

Ethics in AI
The principles and guidelines that govern the responsible development and use of AI technologies. Ethical AI focuses on fairness, transparency, accountability, and ensuring AI does no harm to individuals or society.

Explainable AI (XAI)
Techniques and tools that allow humans to understand how AI models make decisions. XAI is important for increasing transparency and trust, particularly in critical applications like healthcare and finance.

Fine-Tuning

A process in machine learning where a pre-trained model is further trained on a smaller, task-specific dataset. Fine-tuning helps adapt a general-purpose model to a specialized application.

Generative AI

AI models designed to generate new content such as text, images, audio, or video. These models can create outputs that resemble human-made content by learning from vast datasets.

Gradient Descent

An optimization algorithm used to minimize errors in machine learning models by adjusting model parameters (like weights in neural networks). It is essential for training models by finding the best possible performance on a task.

Human-in-the-Loop (HITL)

A process where human judgment is integrated with AI decision-making to improve performance. HITL ensures that human oversight guides AI systems in tasks requiring nuanced understanding or critical outcomes.

Hyperparameters

Settings or configurations that define the structure and learning process of a machine learning model (e.g., learning rate, number of layers). Hyperparameters are tuned to optimize model performance.

Inference

The process of applying a trained AI model to new data to make predictions or decisions. Inference is what happens when an AI model is used in real-world applications after training.

Large Language Models (LLMs)

AI models, like GPT-4, that process and generate human-like text. They are trained on massive datasets of text to understand and predict language patterns, enabling a wide range of applications like answering questions, translation, and content generation.

Loss Function

A mathematical function that quantifies the difference between a model's predictions and the actual outcomes. The goal of training is to minimize the loss function, improving the model's accuracy.

Machine Learning (ML)

A subset of AI where algorithms improve their performance at tasks by learning from data, without being explicitly programmed for each specific task. It involves training models to recognize patterns and make decisions based on data.

Natural Language Processing (NLP)
A field of AI focused on enabling computers to understand, interpret, and generate human language. It encompasses tasks like translation, sentiment analysis, summarization, and conversation generation.

Natural Language Understanding (NLU)
A subfield of NLP focused on enabling machines to comprehend the meaning and intent behind human language. NLU is critical for tasks like question answering, sentiment analysis, and dialogue systems.

Neural Networks
Computational models inspired by the human brain, composed of layers of interconnected nodes ("neurons"). Neural networks are the foundation of most AI models and are used to recognize patterns, classify data, and make decisions.

Nirvana State
A theoretical concept in AI where a system reaches a level of perfection in its performance, free of bias, errors, and inefficiencies. Achieving this state would mean that the AI operates at optimal levels in all aspects, but it is largely seen as an unattainable ideal.

Overfitting
A situation where a machine learning model learns the training data too well, including its noise and errors. Overfitting causes poor performance on new, unseen data, as the model fails to generalize.

Personalization
The use of AI to tailor experiences to individual users, often by analyzing user data and preferences. Personalization is common in recommendation systems (e.g., Netflix, Spotify) and digital marketing.

Prioritization Matrix
A tool used in decision-making to rank AI projects or tasks based on factors such as urgency, impact, and resources. This matrix helps teams focus on high-priority initiatives and allocate resources effectively.

Prompt Engineering
The practice of designing and refining inputs (prompts) for AI models, particularly large language models, to elicit the desired output. It's key to making generative AI models perform specific tasks effectively.

Recurrent Neural Networks (RNNs)
A neural network model designed to process sequential data, such as time series or natural language. RNNs have a memory component that helps them maintain context over sequences, often used in speech recognition and language modeling.

Reinforcement Learning

An area of machine learning where an agent learns to make decisions by interacting with its environment and receiving rewards or penalties based on its actions. This method is often used in gaming, robotics, and complex decision-making scenarios.

Reinforcement Learning from Human Feedback (RLHF)

A type of reinforcement learning where AI models are trained and fine-tuned using human input on their outputs. Humans provide feedback, which the model uses to adjust its actions and improve over time, particularly for tasks requiring nuanced judgment.

Supervised Learning

A machine learning approach where a model is trained on labeled data. The model learns to map inputs to specific outputs by observing examples (e.g., classifying images of cats and dogs).

Training Data

The dataset used to teach an AI model. During training, the model learns from this data to make predictions or decisions. The quality and quantity of training data significantly impact the model's performance.

Transfer Learning

A machine learning technique where a model trained on one task is adapted to perform a different but related task. This allows for faster training and improved performance by leveraging previously learned knowledge.

Transformers

A neural network architecture designed to handle sequential data but more efficiently than RNNs. Transformers are the foundation of many modern AI models, including large language models, by enabling parallel processing of data sequences.

Underfitting

Occurs when a machine learning model is too simple to capture the underlying patterns in the data. An underfitted model performs poorly on both the training data and new data.

Unsupervised Learning

A machine learning technique where models are trained on unlabeled data. The goal is to find hidden patterns or groupings in the data without explicit instructions (e.g., clustering similar items).

DID YOU ENJOY THIS BOOK?

If you enjoyed reading this book, you can help by suggesting it to someone else you think might like it, and **please leave a positive review** wherever you purchased it. This does a lot in helping others find the book. We thank you in advance for taking a few moments to do this.

THANK YOU

You might also like other Thin Leaf Press titles:

Peak Performance: Mindset Tools for Managers
Peak Performance: Mindset Tools for Sales
Peak Performance: Mindset Tools for Leaders
Peak Performance: Mindset Tools for Business
Peak Performance: Mindset Tools for Entrepreneurs
Peak Performance: Mindset Tools for Athletes
The Successful Mind: Tools to Living a Purposeful, Productive, and Happy Life
The Successful Body: Using Fitness, Nutrition, and Mindset to Live Better
The Successful Spirit: Top Performers Share Secrets to a Winning Mindset
Winning Mindset: Elite Strategies for Peak Performance
Winner's Mindset: Peak Performance Strategies for Success
The Life Coach's Tool Kit, Vol. 1
The Life Coach's Tool Kit, Vol. 2
The Life Coach's Tool Kit, Vol. 3
Ordinary to Extraordinary
The Magical Lightness of Being
Explore.

www.ingramcontent.com/pod-product-compliance
Lightning Source LLC
Chambersburg PA
CBHW071320210326
41597CB00015B/1289